STRESS
TRANSIENTS
IN SOLIDS

by

JOHN S. RINEHART

Department of Mechanical Engineering

University of Colorado

and

Technical Director
HyperDynamicS

P.O. BOX 392
SANTA FE, NEW MEXICO 87501

QC
191
.R5

Cover Design: Douglas J. Houston
Graphic Assistant: Rachel Abrams

ISBN No.0-913270-48-2

Library of Congress Catalog Card Number: 75-39138

To my wife.

J.S.R.

PREFACE

It is becoming increasingly important for engineering students and practicing engineers to be know-
ledgeable in the field of impulsive loading: how
stress transients behave in solids. Previous to 1950,
interest in this area was limited principally to the
small group of ordnance engineers concerned with
explosion, fragmentation, and impact. Since then, the
scope and sophistication of the field have enlarged.
In defense, development and use of intercontinental
ballistic missiles have brought problems associated
with the destruction of the missile because of spalling by the stress transients generated through depos-
ition of X ray energy from the blast of an intercept-
ing missile, or the construction of underground
missile storage areas sufficiently hardened to with-
stand the impulsive transient forces produced near
underground or surface bursts of nuclear bombs. The
successful use of nuclear blasts for peaceful purposes with all of the underground nuclear testing
connected with their development, depends critically
on understanding the transient stress effects. The
space age has seen the development of remote control,
a delicate and precise operation most often easily
accomplished with explosives, as well as the protec-
tion of space craft in a hostile environment where
impact with hypervelocity meteoritic material could
be disastrous. Seismological studies have always
involved transient stresses in most of their aspects,
but only recently has it been recognized that transient events have had a major role in producing
important geological features on both the earth and
the moon such as their many impact structures.
    The first book to treat transient stresses at
all fully was STRESS WAVES IN SOLIDS by H. Kolsky,

1953. This was followed shortly, 1954, by my and J. Pearson's BEHAVIOR OF METALS UNDER IMPULSIVE LOADS, a complementary book which treated in detail effects produced in metals by impacts and contact explosions. In 1960, I wrote ON FRACTURES PRODUCED BY EXPLOSIONS AND IMPACTS at the request of the U. S. Air Force to educate some of its personnel in the principles underlying urgent transient stress problems facing them. STRESS WAVE PROPAGATION IN SOLIDS by R. J. Wasley, 1973, discussed some of the newer work, especially elastic-plastic waves. Kolsky's and Wasley's books are still in print.

For some years there has existed a need for a rationally presented, fundamental, yet reasonably elementary discussion of the significant principles controlling the generation, movement, and interaction with boundaries of transient stress waves within solids, such waves as are generated by explosions and impacts. STRESS TRANSIENTS IN SOLIDS satisfies that need. The book has evolved out of twenty years of teaching impulsive loading phenomena to undergraduate and graduate students in mechanical, mining, metallurgical, petroleum, ordnance, geophysical, and geological engineering. Not only is it an appropriate text, but practising engineers in many fields will find it extremely useful where they have occasion to acquire familiarity with transient stress phenomena.

An extreme point of view is adopted here. All bodies are considered to behave in a perfectly linear elastic manner, except for one brief excursion into shock wave phenomena. This attitude has been criticized as being unrealistic but my years of teaching the subject have convinced me that the approach is effective, justified, and useful. It enables the student to appreciate salient aspects, to acquire some skill in dealing with transient stresses, and to solve some very practical problems before encountering the complexities that develop when such concepts as elastic-plastic behavior are introduced. These latter can be built on the firm foundation the student will have acquired.

Many persons have helped me in bringing this book to fruition and to them all my heartfelt thanks. Perhaps my numerous students have aided me most by their generosity and kindness in using portions of the manuscript in classes, checking for errors, and

PREFACE

offering helpful suggestions. I am, however, fully
responsible for what errors remain. My other main
source of inspiration and guidance has been the close
colaborative efforts I have carried on with John
Pearson throughout the past twenty-five years. Several of my colleagues have read and criticized parts
of the manuscript. My wife has encouraged and helped
me always.

May 1975   John S. Rinehart
Santa Fe, New Mexico

iii

CONTENTS

CONTENTS

CHAPTER 1

INTRODUCTION

The simple, yet far reaching relationship which governs the reaction of materials to applied mechanical loads:

Ut tensio sic vis - As the extension, so the force

was first formulated by Robert Hooke in 1660. This theory of elasticity has been continually elaborated with any number of monographs and textbooks discussing at all levels of sophistication problems relating the reaction of elastic bodies to both static and dynamic loading. None has concentrated exclusively on phenomena associated with the passage of short impulsive stress transients through bodies, a field that has now assumed much engineering and geophysical importance. The mechanics and dynamics of the processes leading to fracturing and momentum transfer differ vastly in detail from those which are normally discussed in connection with long duration transients, alternating, and static loading. This book isolates problems peculiar to short transients, discusses many of them in considerable detail, and brings into sharp focus many of the salient aspects of such phenomena. Short lived transient phenomena are considered without regard to the longer lived transient and alternating stresses that they may generate.

The stress transients treated here are ones whose durations are short or commensurate with the time that it takes the stress to traverse the body through which it is traveling. Such stresses which usually arise from impulsive loading play very important roles in determining how bodies break up when struck sharp blows or are subjected to the short intense force of an explosion, and in control-

ling the transfer of momentum along rods and plates
and across boundaries in a wide variety of mechanical and geophysical systems.

The approach is straightforward and slanted
toward obtaining practical solutions to real problems. Except for a single chapter on shock waves,
Chapter 9, all bodies are assumed to behave in a
linear elastic fashion. This much simplified
approach somewhat surprisingly leads to answers that
agree exceedingly well with observed behavior.
Thus the information contained in this book makes
available a number of techniques useful in appre-
ciating the behavior characteristics of materials
and mechanical systems under a wide variety of
transient stress situations.

The developments are based primarily on fundamental physical laws: Hooke's law, which describes
the behavior of elastic bodies to applied forces,
and Newton's laws of motion, force equals mass times
acceleration, and action equals reaction, which relate forces and motions. Knowledge of calculus is
assumed on the part of the reader, but aside from
this, the discussions are complete within themselves. For example, the components of stress and
strain are clearly and fully defined in Chapter 2
and the elastic wave equations are formulated and
the solutions given in Chapter 3. The remaining
chapters are essentially an extensive elaboration
and particularization of these solutions as they
apply to impulsive loading situations.

Readers unfamiliar with transient stress wave
phenomena usually have difficulty grasping the
significant and controlling features which charac-
terize a transient elastic pulse. These features
are discussed in Chapters 4 and 5 for a large
number of specific wave types. The discussion in
Chapter 4 is confined to the simpler plane waves,
with the more complex spherical and cylindrical
waves being dealt with in Chapter 5.

An important and useful extension of Hooke's
law is the principle of superposition which enables
the stresses in interacting waves to be combined.
A number of cases of superposition are explored
in Chapter 6, many of these being situations which
lead to fracturing.

In Chapters 4, 5, and 6, the waves are assumed
traveling in unbounded media. In a real world,
bodies are bounded and the waves will interact

with their boundaries.   These effects are discussed
in detail in Chapter 7.

Most discussions of bar, plate, and surface
waves treat them either as   steady state phenomena
or transient phenomena without any special regard
to their origins. The approach in Chapter 8 is
distinctly different. The dynamics and mechanics
of the conversion of a transient impulsive stress
wave into one of these other types of waves is
followed and emphasized. While much of the discussion is of necessity qualitative, it provides a
physically satisfying bridge between the parent
impulsive load and the waves which it spawns.

Chapter 9, which deals with non Hookean one
dimensional shock wave phenomena, may seem somewhat
out of place. It has seemed, however, important
to include it for the sake of demonstrating and
emphasizing the strong similarities that exist
between the treatment of one dimensional shock wave
phenomena which assumes that the solid behaves as a
perfect fluid, and one dimensional impulsive elastic
stress wave loading phenomena. In many real situations, the transition from one type of behavior to
the other is gradual and subtle, permitting the problem to be treated from a practical viewpoint either
as a shock wave problem or as an elastic wave problem. Because of their great complexities and un-
certainties of applicability, elastic-plastic wave
problems, although of great interest to experimental
and theoretical applied mechanicians, are not
treated here at all.

The details of the processes that govern the
transfer of momentum through a mechanical system
are vastly different when the wave is short or commensurate with the dimensions of the system than when
the duration of the stress is long. The transfer
processes are formulated in Chapter 10 and applied
to a large number of specific simple mechanical
systems.

Beginning with Chapter 11, a number of practical situations are looked at. Most of these involve
interactions between waves which commonly lead to
fracturing. Chapter 11 discusses the techniques,
many closely akin to ones used in optics, for dealing with divergent and convergent waves and nonplanar reflecting and refracting surfaces and interfaces. These nonplanar waves are those most often
produced by explosions and impacts.

The stresses that lead to fracturing fall naturally into two categories. There are those, treated in Chapter 12, which develop when the wave is trapped in a corner and its energy is redistributed as a result of reflections. And then there are those, treated in Chapter 13, in which only a single boundary is involved in the interaction, and spalling occurs.

The intent of this book is to provide a reasonably comprehensive, pedagogically structured discussion of those elements of transient stress waves which can be effectively and broadly applied to the solution of practical problems in the field. While this approach is pragmatic and approximate, it is accurate and useful within the bounds that are prescribed.

CHAPTER 2

COMPONENTS AND MEASURES OF STRESS AND STRAIN

COMPONENTS OF STRESS

The components of stress and strain are defined in this chapter and several useful relationships between them are developed. Every author in beginning his discussion of stresses and strains usually adopts his own favorite notation from the numerous systems in vogue. Here the stresses, and later the strains within a body are defined by considering what is happening across the small, flat elemental plane of surface area δA in Fig. 2-1 with its associated normal OP. The side of the surface in the direction OP is called the "positive side" and that in the opposite direction, the"negative side". The material on one side of the surface exerts a force δF upon the material on the other side. The limit of the ratio δF̄/δA as δA tends to zero is specified as the stress σ̄ at the point O. Stress is a vector with magnitude and dimensions of force per unit area. In general, σ̄ exists for every point in a body, varying from point to point and with time, so that

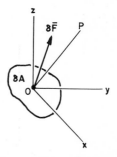

Fig. 2-1. Diagram used to define stress.

$$\bar{\sigma} = \bar{\sigma}(x,y,z,t).$$

Around each point O, the material on the positive side of the surface element δA exerts a force σ̄δA upon the negative side and in accordance with Newton's third law, action equals reaction, the material on the negative side exerts an opposite

5

and equal force on the positive side.

Establishing the coordinate system shown in Fig. 2-2, with the x axis in the direction of OP and δA lying in the yz plane, the x, y, z components of the vector $\bar{\sigma}$ can be written σ_x, τ_{xy}, and τ_{xz}, respectively. The two components, τ_{xy} and τ_{xz}, that lie in the plane δA are designated transverse or shear stresses; σ_x is the normal stress. In this book, σ's will always designate normal stresses

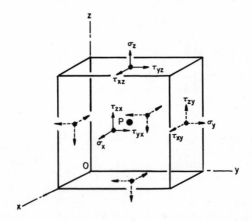

Fig. 2-2. Coordinate system and associated components of normal and shear stress.

and τ's shear stresses. A single subscript on a normal stress indicates the direction in which the stress is acting. The first letter in the shear stress subscript denotes the direction of the stress and the second letter defines the plane in which it is acting. Compressive stresses are considered positive and tensile stresses, negative.

The stresses at O across a plane lying normal to the directions Oy and Oz (Fig. 2-2) are, respectively

$$\tau_{yx}, \ \sigma_y, \ \tau_{yz}$$

and

$$\tau_{zx}, \ \tau_{zy}, \ \sigma_z.$$

Equilibrium considerations require that

$$\tau_{yz} = \tau_{zy}; \quad \tau_{zx} = \tau_{xz}; \quad \tau_{xy} = \tau_{yx}.$$

Fig. 2-2 is often referred to as the elemental stress cube. The six quantities, σ_x, σ_y, σ_z, τ_{xy}, τ_{xz}, and τ_{yz}, specify completely the stress at a point. Each of these quantities is a function of the spatial coordinates, x, y, and z, and time t. When

$$\partial\bar{\sigma}/\partial t = 0$$

for all points, the state of stress is static. Much of the literature in the field of elasticity relates to this type of problem. This book, however, is concerned primarily with dynamic situations where

$$\partial\bar{\sigma}/\partial t \neq 0$$

and the distribution of stress within the body is undergoing both spatial and temporal changes.

It can be shown that it is always possible to make the components of shear on the faces of the elemental cube equal to zero by an appropriate rotation of the coordinate axes. When this is done, the three normal stresses that emerge are called principal stresses, usually designated σ_1, σ_2, and σ_3. A set of principal stresses always exists at each point in a stressed body.

TWO DIMENSIONAL STRESSES

For the special but fairly common two dimensional situation where, with one edge of the elemental reference cube oriented parallel to the z axis,

$$\partial\bar{\sigma}/\partial z = 0$$

one principal stress, σ_3, will lie in the z direction and the other will be given by

$$\sigma_1 = (1/2)(\sigma_x + \sigma_y) + (1/2)\left[(\sigma_x - \sigma_y)^2 + 4\tau_{xy}^2\right]^{\frac{1}{2}} \tag{2.1}$$

and

$$\sigma_2 = (1/2)(\sigma_x + \sigma_y) - (1/2)\left[(\sigma_x - \sigma_y)^2 + 4\tau_{xy}^2\right]^{\frac{1}{2}}. \tag{2.2}$$

The normal stress is a maximum or a minimum and the shear stress is zero on one plane. The plane to which σ_x, σ_y, and τ_{xy} apply intersects this first plane at the angle θ which is given by

$$\tan 2\theta = 2\tau_{xy}/(\sigma_x - \sigma_y). \tag{2.3}$$

A Mohr's circle diagram is a simple and exceedingly useful way of representing stress components for two dimensional situations. The well known construction, illustrated in Fig. 2-3, a and b, makes it possible to determine readily from the known stress components on one plane those on any differently oriented plane, including the principal plane.

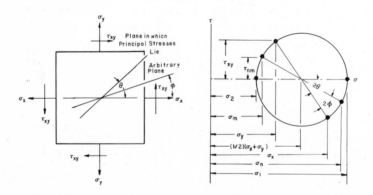

Fig. 2-3.   (a) Notation used in constructing Mohr's circle; (b)   Details of construction.

THREE DIMENSIONAL STRESS

Stress fields within three dimensional bodies must satisfy the conditions of equilibrium. These are obtained by considering the stresses acting on the faces of the elemental cube, $\delta x \delta y \delta z$, shown in

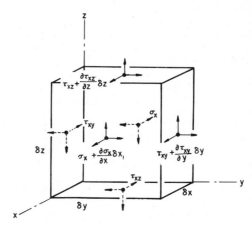

Fig. 2-4. Components of stress acting on the faces of an elemental cube located in a non-uniform stress field.

Fig. 2-4, in which it is assumed that no body forces exist. The sum of the forces in the x direction are

$$\left[\sigma_x + (\partial\sigma_x/\partial x)\delta x\right]\delta y \delta z - \sigma_x \delta y \delta z +$$

$$+ \left[\tau_{xy} + (\partial\tau_{xy}/\partial y)\delta y\right]\delta x \delta z - \tau_{xy}\delta x \delta z + \qquad (2.4)$$

$$+ \left[\tau_{zx} + (\partial\tau_{zx}/\partial z)\delta z\right]\delta x \delta y - \tau_{zx}\delta x \delta y = 0$$

Similar sums can be written for the resultant forces in the y and z directions. The sums reduce to the three conditions of equilibrium

$$\partial\sigma_x/\partial x + \partial\tau_{xy}/\partial y + \partial\tau_{zx}/\partial z = 0$$

$$\partial\sigma_y/\partial y + \partial\tau_{yz}/\partial z + \partial\tau_{xy}/\partial x = 0 \qquad (2.5)$$

$$\partial\sigma_z/\partial z + \partial\tau_{zx}/\partial x + \partial\tau_{yz}/\partial y = 0.$$

These equations, the compatability equations, and the influence of constraints imposed by boundaries are sufficient to describe how a stressed body deforms.

STRESS-STRAIN RELATIONSHIPS

For small deformations most materials are elastic, the deformations being reversible without any time delay, and obey Hooke's law. First formulated by Robert Hooke in 1660, this law states that stress and strain are linearly related. To the degree that this relationship holds, superposition of stresses or deformations is simply a matter of linear addition of components.

In elastic theory, deformations or strains are described in terms of changes in length and changes in the angle between two lines or between a line and a plane. The first is called an extension and the second a shear. If s is the distance between two near points O and P, and s + δs between the corresponding points O' and P' after straining, then ε, the linear extension strain, is defined as the change in length divided by the original length, or

$$\varepsilon = \delta s/s.$$

Now consider two lines OP and OR, perpendicular at O in the unstrained state, that are strained to the configuration shown in Fig. 2-5. The shear

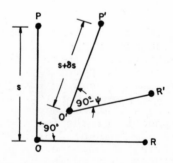

Fig. 2-5. Diagram used to define strain.

strain γ associated with these directions is defined
as

$$\gamma = \tan \psi.$$

where ψ is the angle through which the lines OP and
OR have rotated to their new positions O'P' and
O'R'.   The usual assumption is that the quantities
ε and γ are so small that their squares are
negligible.
     Customarily the displacements of particles of a
deformed body are resolved into the components u, v,
and w parallel to the coordinate axes x, y, and z,
respectively. The nine quantities

$$\partial u/\partial x, \quad \partial u/\partial y, \quad \partial u/\partial z, \quad \partial v/\partial x, \quad \partial v/\partial y,$$

$$\partial v/\partial z, \quad \partial w/\partial x, \quad \partial w/\partial y, \quad \partial w/\partial z$$

which are functions of x, y, z, and t, define the
strain at a point. If their values are known at
this point, the relative displacement of all surrounding points may be found. For convenience,
these nine quantities are regrouped and denoted as
follows:

$$\varepsilon_x = \partial u/\partial x \quad \gamma_{xy} = \partial u/\partial y + \partial v/\partial x \quad 2\bar{\omega}_x = \partial w/\partial y - \partial v/\partial z$$

$$\varepsilon_y = \partial v/\partial y \quad \gamma_{xz} = \partial u/\partial z + \partial w/\partial x \quad 2\bar{\omega}_y = \partial u/\partial z - \partial w/\partial x$$

$$\varepsilon_z = \partial w/\partial z \quad \gamma_{yz} = \partial v/\partial z + \partial w/\partial y \quad 2\bar{\omega}_z = \partial v/\partial x - \partial u/\partial y.$$
$$(2.6)$$

     The quantities ε_x, ε_y, and ε_z are fractional
expansions and contractions of infinitesimal line
elements passing through P (Fig. 2-2) parallel to
the x, y, and z axes, respectively. The quantities
γ_{xy}, γ_{xz}, and γ_{yz} correspond to the components of
shear strain in the planes denoted by their suffixes.
The quantities $\bar{\omega}_x$, $\bar{\omega}_y$, and $\bar{\omega}_z$ are not deformations
but are the components of the rotation of the ele-
mental cube (Fig. 2-2) about P as a rigid body.
     If the material is homogeneous and isotropic
and reacts to stress in a perfectly elastic and
linear manner, two elastic constants are sufficient
to express the relationships between stress and
strain. When the material is totally anisotropic,

twenty-one elastic constants are needed. A convenient and frequently used pair of elastic constants
for isotropic materials are Lamé's parameters, λ
and G. These relate the principal stress and strain
components through the following equations:

$$\sigma_1 = (\lambda + 2G)\varepsilon_1 + \lambda\varepsilon_2 + \lambda\varepsilon_3$$

$$\sigma_2 = \lambda\varepsilon_1 + (\lambda + 2G)\varepsilon_2 + \lambda\varepsilon_3 \qquad (2.7)$$

$$\sigma_3 = \lambda\varepsilon_1 + \lambda\varepsilon_2 + (\lambda + 2G)\varepsilon_3 .$$

The convenience of using Lamé's parameters, particularly in wave propagation problems, is that the
combination $\lambda + 2G$ relates stress and strain in the
same direction and λ relates them in the perpendicular directions.

The sum of the three principal strains, $\varepsilon_1 +
\varepsilon_2 + \varepsilon_3$, neglecting second order terms, is the frac-
tional volume change or dilatation Δ so that the
above equations can be written:

$$\sigma_1 = \lambda\Delta + 2G\varepsilon_1 \qquad \sigma_2 = \lambda\Delta + 2G\varepsilon_2$$

$$\sigma_3 = \lambda\Delta + 2G\varepsilon_3 . \qquad (2.8)$$

For cartesian coordinates, the complete set of
equations connecting stress and strain are

$$\sigma_x = \lambda\Delta + 2G\varepsilon_x \qquad \tau_{yz} = G\gamma_{yz}$$

$$\sigma_y = \lambda\Delta + 2G\varepsilon_y \qquad \tau_{zx} = G\gamma_{zx} \qquad (2.9)$$

$$\sigma_z = \lambda\Delta + 2G\varepsilon_z \qquad \tau_{xy} = G\gamma_{xy}.$$

Other commonly used equivalent elastic constants
are the modulus of rigidity, the ratio of shear
stress to shear strain in simple shear, which is just
the Lamé parameter G; the bulk modulus or incom-
pressibility K, the ratio of hydrostatic pressure to
the dilatation it produces; Young's modulus, E, the
ratio of tension to extension in a cylinder which is
under axial tension and which is unrestricted laterally, the case of plane stress; and Poisson's ratio,
ν, the ratio of lateral contraction to longitudinal
extension for the above cylinder.

Under a hydrostatically applied pressure, P

$$\sigma_1 = \sigma_2 = \sigma_3 = p$$

so that

$$\sigma_1 + \sigma_2 + \sigma_3 = 3p$$

which from Eq. (2.8) gives

$$3\lambda\Delta + 2G(\varepsilon_1 + \varepsilon_2 + \varepsilon_3) = 3p$$

giving

$$K = p/\Delta = \left[\lambda + (2/3)G\right] . \qquad (2.10)$$

The reciprocal of K is called the compressibility.
From the definition of Young's modulus,
$\sigma_2 = \sigma_3 = 0$, so that Eq. (2.8) becomes

$$\sigma_1 = (\lambda + 2G)\varepsilon_1 + \lambda\varepsilon_2 + \lambda\varepsilon_3$$

$$0 = \lambda\varepsilon_1 + (\lambda + 2G)\varepsilon_2 + \lambda\varepsilon_3$$

$$0 = \lambda\varepsilon_1 + \lambda\varepsilon_2 + (\lambda + 2G)\varepsilon_3$$

giving

$$\varepsilon_2 = \varepsilon_3 = -\left[\lambda/2(\lambda + G)\right]\varepsilon_1$$

and

$$E = \sigma_1/\varepsilon_1 = G(3\lambda + 2G)/(\lambda + G) . \qquad (2.11)$$

Poisson's ratio is related to λ and G by the
equation

$$\nu = \lambda/2(\lambda + G). \qquad (2.12)$$

There are many relationships among the several
elastic constants, λ, G, K, E, and ν. The following
are frequently useful:

$$\lambda = E\nu/(1 + \nu)(1 - 2\nu) = K - (2/3)G = 2\nu G/(1-2\nu)$$

$$K = 2(1 + \nu)G/3(1 - 2\nu) = E/3(1 - 2\nu); \quad G = E/2(1 + \nu)$$

$$E = 9KG/(3K + G); \quad \nu = (3K - 2G)/2(3K +G).$$

$$(2.13)$$

In the treatment of plane wave problems, the axes can be appropriately oriented so that the problem becomes essentially two dimensional. For such a wave in a medium of infinite extent $\varepsilon_1 \neq 0$ and $\varepsilon_2 = \varepsilon_3 = 0$, and Eq. (2.8) becomes

$$\sigma_1 = (\lambda + 2G)\varepsilon_1$$

$$\sigma_2 = \lambda\varepsilon_1$$

$$\sigma_3 = \lambda\varepsilon_1$$

so that

$$\sigma_2 = \sigma_3 = [\lambda/(\lambda + 2G)]\sigma_1 = [\nu/(1 - \nu)]\sigma_1. \qquad (2.14)$$

The strains are explicitly related to the stresses by the following equations derived from Eq. (2.9) and the first two equations of Eq. (2.13):

$$\varepsilon_x = [\sigma_x - \nu(\sigma_y + \sigma_z)]/E \qquad \gamma_{yz} = 2(1 + \nu)\tau_{yz}/E$$

$$\varepsilon_y = [\sigma_y - \nu(\sigma_x + \sigma_z)]/E \qquad \gamma_{zx} = 2(1 + \nu)\tau_{zx}/E$$

$$\varepsilon_z = [\sigma_z - \nu(\sigma_x + \sigma_y)]/E \qquad \gamma_{xy} = 2(1 + \nu)\tau_{xy}/E.$$

$$(2.15)$$

STRESS-STRAIN IN NONCARTESIAN COORDINATES

Waves emanating from impacts and explosions frequently possess cylindrical or spherical symmetry, making it convenient to express the stress and strain relationships in terms of cylindrical or spherical coordinates.

For a cylindrically symmetric situation, the equations for the radial and circumferential or tangential stresses can be written in the form

$$\sigma_r = (\lambda + 2G)r^{-1}\partial(ru)/\partial r - 2Gu/r$$

$$\sigma_\theta = \lambda r^{-1}\partial(ru)/\partial r + 2Gu/r \qquad (2.16)$$

$$\sigma_z = \lambda r^{-1}\partial(ru)/\partial r.$$

Combining the above equations gives

$$\sigma_r + \sigma_\theta = [2(\lambda + G)/\lambda]\sigma_z. \qquad (2.17)$$

For a spherically symmetric situation, also of considerable interest in connection with the propagation of impulsively generated waves by explosion or impact, the equations for the principal stresses can be written in the form

$$\sigma_r = (\lambda + 2G)\partial u/\partial r + 2\lambda u/r$$
$$\sigma_\theta = \lambda\partial u/\partial r + 2(\lambda + G)u/r. \qquad (2.18)$$

Dilatation, $\bar{\nabla}\cdot\bar{u}$. for the spherically symmetric case takes the form

$$\bar{\nabla}\cdot\bar{u} = r^{-2}\partial(r^2 u)/\partial r \qquad (2.19)$$

giving, on combination of Eqs. (2.18) and (2.19)

$$\sigma_r = (\lambda + 2G)\bar{\nabla}\cdot\bar{u} - 4Gu/r$$
$$\sigma_\theta = \lambda\bar{\nabla}\cdot\bar{u} + 2Gu/r. \qquad (2.20)$$

These many relationships will prove useful later on in deriving equations describing stresses and strains associated with the propagation of plane, cylindrical, and spherical transient elastic waves.

VALUES OF ELASTIC PARAMETERS

For many solids, the Lamé constant λ is very nearly equal to the shear modulus G, implying a Poisson's ratio of 0.25, and greatly simplifying calculations. This assumption makes

$$K = 5G/3, \ E = 5G/2.$$

The Poisson's raios of most solids lie in the range from 0.2 to 0.4.

Strain is a dimensionless ratio although occasionally referred to in such units as inches per inch or centimeters per centimeter. Since stress has the dimensions of force per unit area, all of the elastic moduli and constants do likewise. Various stress units in common usage are dynes per cm^2, lb per in^2, kilogram per cm^2, atmosphere, bars, and kilobars. Relationships among the various stress units are tabulated in Table 2-1.

The values of the elastic moduli vary from material to material. They are also influenced, often strongly, by ambient environmental conditions such as temperature and pressure, and by previous mechanical working. Some typical values are listed in Table 2-2.

Table 2-1.   Stress unit equivalents

$$1 \; dyne/cm^2 = 1.45 \times 10^{-5} \; lb/in^2$$
$$= 1.02 \times 10^{-6} \; kg/cm^2$$
$$= 1 \; barye$$

1033×980 barye = 1 atmos = 1.013×10^6 barye = 1.013 bars

1 newton = 1 kg/m^2

standard atmos = 76 cm/Hg = 30 in Hg = 34 ft water
$$= 14.7 \; lb/in^2 = 1033 \; grams/cm^2$$
$$= 1.013 \times 10^6 \; dyne/cm^2$$

1 kilobar = 10^3 bars $\simeq$ 15,000 lb/in^2

Table 2-2.   Typical values of elastic moduli

Material	Modulus - $10^{11} dyne/cm^2$			
	E	G	K	ν
Steel	20.9	8.1	16.6	0.29
Copper	12.3	4.5	15.4	0.37
Lead	1.6	0.56	3.8	0.43
Quartz fiber	5.2	3.0	1.4	-
Rubber	0.05	-	-	0.46
Granite*	4.6	1.9	2.6	0.21
Limestone*	5.8	2.3	4.0	0.26
Sandstone*	5.7	2.6	2.3	0.10
Water			0.2	

*Elastic properties of rocks of the same
general type vary greatly

Chapter 3

FUNDAMENTAL ELASTIC WAVE EQUATIONS

DERIVATIONS OF WAVE EQUATIONS

When an elastic body is disturbed by suddenly displacing a portion of it, some time usually elapses before the remainder of the body is affected by the displacement. Inertial and elastic prop- erties control the velocity of the advance of the disturbance: the greater the density of a body, the lower the velocity. Velocity is higher the more resilient a body is. The form of the disturbance frequently changes as it progresses from point to point, depending upon the initial character of the displacement and upon any boundaries that it may encounter as it traverses the body.

The equations of motion of the disturbance can be obtained by considering the variations in stress occurring across the elemental cube illustrated in Fig. 2-4 as a function of time. The resultant force in the x direction, from Eq. (2.5) is

$$(\partial \sigma_x / \partial x + \partial \tau_{xy} / \partial y + \partial \tau_{xz} / \partial z) \delta x \delta y \delta z.$$

Neglecting body forces such as gravity, by Newton's second law of motion, the force in the x direction will be equal to $[(\rho \delta x \delta y \delta z)(\partial^2 u / \partial t^2)]$ where ρ is the density of the element. Equating the two forces yields

$$\rho(\partial^2 u / \partial t^2) = \partial \sigma_x / \partial x + \partial \tau_{xy} / \partial y + \partial \tau_{xz} / \partial z. \qquad (3.1a)$$

Similarly

$$\rho(\partial^2 v / \partial t^2) = \partial \tau_{yx} / \partial x + \partial \sigma_y / \partial y + \partial \tau_{yz} / \partial z \qquad (3.1b)$$

17

and

$$\rho(\partial^2 w/\partial t^2) = \partial\tau_{zx}/\partial x + \partial\tau_{zy}/\partial y + \partial\sigma_z/\partial z. \qquad (3.1c)$$

These equations are valid whatever the stress-strain behavior of the medium. Only the elastic case is considered here. For this case, by using the stress-strain relationships given in Eq. (2.9), the first of the above equations can be written

$$\rho(\partial^2 u/\partial t^2) = \partial(\lambda\Delta+2G\varepsilon_x)/\partial x+\partial(G\gamma_{xy})/\partial y+\partial(G\gamma_{xz})/\partial z. \qquad (3.2)$$

From the defining equation, Eq. (2.6)

$$\varepsilon_x = \partial u/\partial x; \quad \gamma_{xz} = \partial u/\partial z+\partial w/\partial x; \quad \gamma_{xy} = \partial u/\partial y+\partial v/\partial x$$

substitution in Eq. (3.2) yields

$$\rho\partial^2 u/\partial t^2 = (\lambda + G)\partial\Delta/\partial x + G\nabla^2 u \qquad (3.3a)$$

where ∇^2 is the operator

$$(\partial^2/\partial x^2 + \partial^2/\partial y^2 + \partial^2/\partial z^2).$$

Similarly

$$\rho\partial^2 v/\partial t^2 = (\lambda + G)\partial\Delta/\partial y + G\nabla^2 v \qquad (3.3b)$$

and

$$\rho\partial^2 w/\partial t^2 = (\lambda + G)\partial\Delta/\partial z + G\nabla^2 w. \qquad (3.3c)$$

Eqs. (3.3a), (3.3b), and (3.3c) are the equations of motion that govern the propagation of transient disturbances through isotropic elastic bodies. It will be shown below that two and only two modes of propagation are permitted, distortion and dilatation, each traveling with its own characteristic velocity.

Differentiating both sides of Eq. (3.3a) with respect to x, both sides of Eq. (3.3b) with respect to y, and both sides of Eq. (3.3c) with respect to z, and adding, yields the wave equation

$$\rho\partial^2\Delta/\partial t^2 = (\lambda + 2G)\nabla^2\Delta. \qquad (3.4)$$

The dilatation Δ is thus propagated through the body

with the velocity $[(\lambda + 2G)/\rho]^{\frac{1}{2}}$.

Now eliminating Δ between Eq. (3.3b) and Eq. (3.3c) by differentiating both sides of Eq. (3.3b) with respect to z and Eq. (3.3c) with respect to y, and subtracting, yields

$$\rho \partial^2 (\partial w/\partial y - \partial v/\partial z)/\partial t^2 = G\nabla^2 (\partial w/\partial y - \partial v/\partial z)$$

or

$$\rho \partial^2 \bar{\omega}_x/\partial t^2 = G\nabla^2 \bar{\omega}_x \qquad (3.5)$$

where from Eq. (2.6), $\bar{\omega}_x$ is the rotation about the x axis. Similar equations hold for $\bar{\omega}_y$ and $\bar{\omega}_z$. The rotation is thus propagated with velocity $(G/\rho)^{\frac{1}{2}}$.

For zero dilatation, Eq. (3.3a) becomes

$$\rho \partial^2 u/\partial t^2 = G\nabla^2 u \qquad (3.6)$$

with similar equations for v and w. The rotations, $\bar{\omega}_x$, $\bar{\omega}_y$, and $\bar{\omega}_z$ vanish if u, v, and w are derivable from a potential function ϕ as follows:

$$u = \partial\phi/\partial x, \quad v = \partial\phi/\partial y, \quad w = \partial\phi/\partial z$$

so that

$$\Delta = \nabla^2 \phi, \quad \partial\Delta/\partial x = \nabla^2 u.$$

Substitution in Eq. (3.6) gives

$$\rho \partial^2 u/\partial t^2 = (\lambda + 2G)\nabla^2 u \qquad (3.7)$$

and similarly for v and w.

Eqs. (3.6) and (3.7) indicate that solid waves can propagate through an elastic solid with two different velocities. Those involving no rotation, irrotational waves, propagate with velocity

$$[(\lambda + 2G)/\rho]^{\frac{1}{2}} = c_1.$$

and those involving no dilatation, equivoluminal waves with velocity c_1, given by $(G/\rho)^{\frac{1}{2}}$. The irrotational waves are usually called dilatation or longitudinal waves, and the equivoluminal waves are called shear, transverse, or distortion waves. The term distortion wave is somewhat misleading in that irrotational waves involve both dilatation and distortion but equivoluminal waves involve only distortion.

    A plane wave must, in fact, travel in an isotropic elastic medium with one or the other of the above velocities. Since the material is isotropic, the plane wave can be considered to be propagating along the x axis without any loss of generality. The displacements u, v, and w will be functions of a single parameter, $\Psi = (x - ct)$ where c is the velocity of propagation of the wave. Since the wave is plane, it propagates without change in form. At point x, at distance ct along the x axis, a point in the body will experience at time t_1 exactly the same displacements as a point x_2 at distance ct_2 along the x axis will experience at time t_2, t_1 and t_2 having a common time base.

    Partial differentiation of the function $u(\Psi)$, $v(\Psi)$, and $w(\Psi)$ yields the following relationships:

$$\partial^2 u/\partial t^2 = c^2 \partial^2 u/\partial \Psi^2 \qquad \partial^2 u/\partial x^2 = \partial^2 u/\partial \Psi^2$$

$$\partial^2 v/\partial t^2 = c^2 \partial^2 v/\partial \Psi^2 \qquad \partial^2 v/\partial x^2 = \partial^2 v/\partial \Psi^2 \qquad (3.8)$$

$$\partial^2 w/\partial t^2 = c^2 \partial^2 w/\partial \Psi^2 \qquad \partial^2 w/\partial x^2 = \partial^2 w/\partial \Psi^2.$$

The differentials with respect to y and z are all zero. Substitution of the right hand differentials with respect to Ψ into Eq. (3.6) yields

$$\rho c^2 \partial^2 u/\partial \Psi^2 = (\lambda + 2G)\partial^2 u/\partial \Psi^2. \qquad (3.9)$$

Similarly

$$\rho c^2 \partial^2 v/\partial \Psi^2 = G \partial^2 v/\partial \Psi^2 \qquad (3.10)$$

and

$$\rho c^2 \partial^2 w/\partial \Psi^2 = G \partial^2 w/\partial \Psi^2. \qquad (3.11)$$

Eqs. (3.9), (3.10), and (3.11) can be satisfied only in two ways. Either

$$c^2 = (\lambda + 2G)/\rho$$

and $\partial^2 v/\partial \Psi^2$ and $\partial^2 w/\partial \Psi^2$ are equal to zero, or

$$c^2 = G/\rho$$

and $\partial^2 u/\partial\Psi^2$ is equal to zero. A longitudinal wave, with motion in the direction of propagation of the wave will satisfy the first set of conditions. The second set of conditions is satisfied by a wave moving in the x direction but with displacements restricted to transverse motions lying in the plane of the wave front.

Even though the direction of displacement in a longitudinal wave is restricted to the direction of propagation of the wave, both the bulk modulus and the shear modulus control the velocity of propagation, the velocity in terms of these moduli being given by

$$\left\{[K + (4/3)G]/\rho\right\}^{\frac{1}{2}}.$$

The reason for this is illustrated in Fig. 3-1 where the dilatation wave during passage alters the shape of a small cube of material. Its x dimension

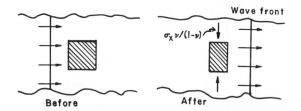

Fig. 3-1. Change in shape of an elemental cube produced by engulfment in a plane transient stress wave.

is shortened while its y and z dimensions remain the same. Changes in shape involve the resistance of the material to both shear and compression. The strain in a direction, for example y, parallel to the front is zero so that from Eq. (2.15)

$$\varepsilon_y = 0 = \sigma_y/E - \nu(\sigma_x + \sigma_y)/E,$$

yielding the following relationship between σ_x and σ_y:

$$\sigma_y = \sigma_x\nu/(1 - \nu). \qquad (3.12)$$

Since the direction is arbitrary, except that it must lie in the plane of the wave front, the stress

in any direction parallel to the wave front,will be
$\nu/(1 - \nu)$ times the stress in the wave.

SOLUTION OF THE WAVE EQUATION FOR A PLANE WAVE

The governing wave equation, Eq. (3.6), is of
the form

$$\partial^2\alpha/\partial t^2 = c^2\nabla^2\alpha \qquad\qquad (3.13)$$

which, for a plane wave traveling in the x direction,
becomes

$$\partial^2\alpha/\partial t^2 = c^2\partial^2\alpha/\partial x^2. \qquad\qquad (3.14)$$

The general solution of this equation is

$$\alpha = f(x - ct) + F(x + ct) \qquad\qquad (3.15)$$

f and F being arbitrary functions that depend on
initial conditions. The function f represents a
plane wave traveling in the positive x direction,
and F a wave in the opposite direction.

SOLUTION OF THE WAVE EQUATION FOR NONPLANAR
WAVES

Many disturbances give rise to wave fronts that
are curved, both convergent and divergent. Two cases
of particular importance are spherical expanding
and converging waves. These are not steady state
situations as is the case with a plane wave in which
the distribution of energy along the wave is fixed
as the wave propagates so that the wave maintains
a constant shape. The energy along a curved wave
front is being constantly redistributed over the
changing area of the front, causing progressive
changes in the distribution of energy in that part
of the wave lying behing the front.
Specific solutions have been obtained for a
number of cases and these will be discussed in Chapter 5. Each solution depends strongly on initial
conditions.

VALUES OF WAVE VELOCITIES

In general, the two basic wave velocities, the dilatation wave velocity, c_1, and the shear wave velocity, c_2, are of the order of a few thousand meters per second. Such a wave will move through a distance of one centimeter in about two microseconds. The shear wave velocity is always less than the dilatation wave velocity, usually about one half. If a sudden deformation generates both types, the longitudinal wave rapidly outruns the shear wave and the two become more and more separated.

Typical values of the velocities c_1 and c_2 are listed in Table 3-1 for a number of materials.

Table 3-1.   Typical dilatation and shear wave velocities

Material	c_1 (m/sec)	c_2 (m/sec)
Aluminum	6,100	3,100
Brass	4,300	2,000
Glass (window)	6,800	3,300
Steel	5,800	3,100
Lead	2,200	700
Plexiglas	2,600	1,300
Polystyrene	2,300	1,200
Magnesium	6,400	3,100

These are only representative values since they are sensitive to state of stress, temperature, composition, mechanical history, and mechanical state of each material. Generally in metals, wave velocities are relatively insensitive to these factors, it being necessary to reach high temperatures or high pressures to produce significant changes. But in rocks these same factors strongly influence wave velocities by consolidating the rock, closing microfractures and voids, and increasing the resilience, thus enabling the rock to transmit energy faster. As an example, the effect of moderate pressure on wave propagation velocity is illustrated in Fig. 3-2 for dolomite. The composition and structure varies greatly from rock to rock in a given rock type and

causes a wide spread in the values of their dilatation wave velocities.

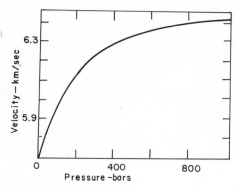

Fig. 3-2. Dilatational velocity in dolomite at
27°C as a function of confining pressure.

Many schemes have been developed for measuring
c_1 and c_2 both in the laboratory and in the field.
By determining both quantities in any particular
material, assuming it is isotropic, it is possible
to calculate all of the elastic parameters of the
material, utilizing, initially, the relations

$$c_1 = \left[(\lambda + 2G)/\rho\right]^{\frac{1}{2}} \qquad (3.16)$$

and

$$c_2 = (G/\rho)^{\frac{1}{2}} \qquad (3.17)$$

to calculate λ and G, and then Eq. (2.13) to calculate E, K, and ν. Poisson's ratio is simply related
to the two velocities through the relationship

$$c_1/c_2 = \left[2(1 - \nu)/(1 - 2\nu)\right]^{\frac{1}{2}}. \qquad (3.18)$$

Chapter 4

CHARACTERISTICS OF TRANSIENT ELASTIC PLANE WAVES IN UNBOUNDED MEDIA

SOME SIMPLE SOLUTIONS

A disturbance, initially starting from some region, may have one of many forms when it finally arrives in the region where it is being observed. When the region of observation is at a great distance from the region of initiation, the radius of curvature of the wave front will become infinite so that the front can be treated as plane. Only plane waves are considered in this chapter; nonplanar waves behave quite differently and are discussed in Chapter 5.

To illustrate the main features of plane elastic waves in an unbounded solid, first assume that the displacement is parallel to the x axis. The y and z displacements, v and w respectively, are both zero. And second, assume the x displacement, u, is independent of y and z. These two assumptions specify the conditions that define a longitudinal plane wave. Eq. (3.13) becomes

$$\delta^2\alpha/\delta t^2 = c_1^2 \delta^2\alpha/\delta x^2$$

where

$$c_1 = \left[(\lambda + 2G)/\rho\right]^{\frac{1}{2}}.$$

The general solution of this is

$$\alpha(x,t) = f(x - c_1 t) + F(x + c_1 t). \qquad (4.1)$$

Either of the two right hand terms describes an arbitrary wave form, the first, f, a wave propagating

25

in the positive x direction, and the second, F, in the negative x direction.

There are two constructive and useful ways to look at the disturbance associated with the passage of an elastic wave. One is to observe the behavior of a single point (x = constant) as a function of time. The other is to examine the behavior of the region in the neighborhood of a single point at any particular time (t = constant). Both approaches are used here.

For a wave whose motion is in the positive x direction

$$\alpha(x,t) = f(x - ct).$$

For a longitudinal wave, the stress σ_x will be given by

$$\sigma_x = f(x - c_1 t) \qquad (4.2)$$

and for a shear wave

$$\tau_{yz} = f(x - c_2 t) \qquad (4.3)$$

where f is any function of (x - ct).

The form of a transient wave depends upon the source of the disturbance, the distance of travel, and the material. Generally, impacts and explosions produce disturbances that appear suddenly, persist for a short time. and then die away. A wide variety of wave forms is possible such as those illustrated in Fig. 4-1. The square pulse, Type a, holds a special place since for computational purposes any of the other forms can be constructed by grouping together square pulses of varying heights and arbitrarily chosen narrownesses. Each of these types can be expressed analytically in a fairly simple fashion.

As long as the transient wave is elastic and plane, its shape does not change as it moves through a material. Every part of the wave is propagated with the same velocity, c_1, if it is a dilatation wave and c_2 if it is a distortion wave. Such a propagating wave of sawtooth shape is illustrated in Fig. 4-2. The deformation at the point a' at the time t_2 is exactly the same as it was at the point a at time t_1, and the deformation at the point b' at time t_2 is exactly the same as it was at the point b at time t_1, and so on, both a and a', b and b', and c and c' being separated by a distance $c(t_2 - t_2)$.

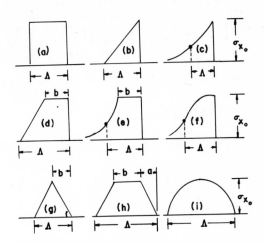

Fig. 4-1.   Variety of transient wave forms. Ordinate
             is compressive stress and abcissa is
             distance.   (a) square pulse; (b) saw-
             tooth wave; (c) exponentially decaying
             wave; (d) flattopped wave; (e) flattopped
             exponentially decaying wave; (f) bell
             shaped wave; (g) triangular wave; (h) fast
             rise, slow decay wave; (i) half sine wave.

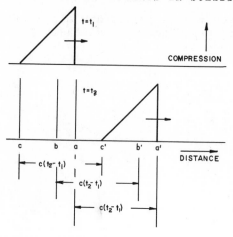

Fig. 4-2. Stress states associated with sawtooth compression wave advancing to the right with velocity c.

DEPICTION OF PROGRESS OF A WAVE

A common way of depicting the movement of a plane elastic wave is to plot the rate of progress of a given state of deformation, placing distance along the abcissa and time along the ordinate. Such plots, called Lagrangian diagrams, are especially useful in describing reflection and interference of transient waves. Figure 4-3 illustrates the progress of the sawtooth wave drawn in Fig. 4-2. Here each

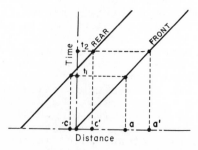

Fig. 4-3. Position of sawtooth wave of Fig. 4-2 as function of time.

line corresponds to a particular state of deformation. The slope of each line is $1/c$.

Figure 4-4 illustrates the progressive separation of a distortion wave front and a dilatation
wave front with both originated at the same place
at time zero but then separated because of their
different respective velocities. The slope of the
time-distance line of the distortion wave is $1/c_2$
whereas that of the dilatation wave is $1/c_1$.

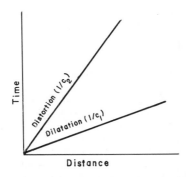

Fig. 4-4.   Relative positions of distortion wave
            front and dilatation wave front as
            function of time, both originating at
            same point.

PARTICLE VELOCITY, THE CONCEPT

During the passage of a transient stress wave,
the ever changing nature of the deformations is
accomplished through movement within the material.
At times it is important and useful to describe
these movements quantitatively and in detail, relating them to associated stresses and accompanying
deformations. The instantaneous velocity v of a
moving point within the medium is called the
particle velocity of that point. A particle velocity
can be assigned to every point of a disturbed
medium. In an elastic body transmitting a wave,
the particle velocity at any point in the wave is
linearly related to the instantaneous stress at
that point.

Consider a portion of the plane transient dilatation wave in Fig. 4-5 which at time $t = t_1$ is coincident with the reference line MN. This might, for example, be the point b of the sawtooth wave shown in Fig. 4-2. Shortly thereafter at $t = t_1 + \Delta t$, this part of the wave will have moved

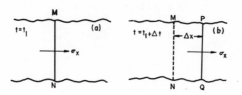

Fig. 4-5.   Momentum transfer during progress of transient stress wave.

the distance Δx beyond MN to a new position, PQ in Fig. 4-5b. During this time, the stress σ_x, corresponding to the stress level for that portion of the wave under consideration, will have been acting on the material enclosed by the reference lines MN and PQ, imparting to this material the impulse $\sigma_x \Delta t$. Since the mass of material per unit cross-sectional area enclosed between the two reference lines is equal to $\rho \Delta x$, application of Newton's second law of motion, impulse is equal to change of momentum, gives the following relationship

$$\sigma_x \Delta t = \rho v_x \Delta x = \rho v_x c_1 \Delta t$$

which reduces to

$$\sigma_x = \rho c_1 v_x \qquad (4.4)$$

where v_x is particle velocity directed along the direction of propagation of the wave. Similarly it can be shown that for a distortion wave

$$\tau_{yz} = \rho c_2 v_y \qquad (4.5)$$

where the particle velocity, v_y, is now perpendicular to the direction of propagation of the wave. It must be emphasized that these relationships are valid only for a single wave and care must be exercised when two or more waves are superposed. Thus stress and

particle velocity are linearly related, the proportionality constant being the product, ρc,
density times wave velocity, commonly called the
specific acoustic resistance of the material.
  As mentioned above, two points of view are
usually employed in viewing transient wave motion:
in one, the wave is viewed as a spatial variation
of stress or particle velocity; and in the other,
the stress at, or the motion of, a particular point
is examined as a function of time. The equation
relating stress, wave velocity, and particle velocity
makes it easy to shift from one point of view to
another. The first point of view was illustrated in
Fig. 4-2 where except for conversion factors and
units, the sawtooth wave would have exactly the same
shape in (σ, t), (v, t), (v, x), and (σ, x) plots.
  The ratio of σ/v, which is equal to the specific
acoustic resistance, the product ρc, relates stress
and particle velocity directly. For most materials,
as is evident in Table 4-1, the product ρc for
dilatation waves lies in the range from 0.3×10^4
to 4.5×10^4 gm/sec-cm. For steel, a stress of about
0.14 kg/cm^2 corresponds to a particle velocity of
about 30 cm/sec. High stresses, of the order of
300 kilobars, such as are generated by explosive
charges detonated against steel plates, develop
particle velocities lying in the neighborhood of
800 m/sec.

Table 4-1. Values of specific acoustic resistance

Material	c_1 (m/sec)	ρ (gm/cm^3)	ρc_1 (σ/v) (gm/sec-cm^2 x 10^4)
Aluminum	6100	2.7	165
Brass	4300	8.4	361
Plexiglas	2600	1.2	31
Steel	5800	7.8	452
Lead	2200	11.3	249

DISPLACEMENT

A common consequence of the passage of a transient stress pulse through a material is the permanent displacement of the material, contrasting markedly with the to and fro movement about an equilibrium position excited by steady state oscillations. The total distance, d, that a point subjected to the action of a transient longitudinal stress wave $\sigma_x(t)$ will be transported is given by the equivalent relationships

$$d = \int_0^T v(t)\,dt = (1/\rho c)\int_0^T \sigma_x(t)\,dt \qquad (4.6)$$

where $v(t)$ is the distribution of particle velocities in the transient wave and T is its duration. Generally when a transient wave arrives at a point within a body, the point suddenly begins to move, moves for a while, and then comes to rest in an unstressed state, there being no residual relative displacment between adjacent points. During this time, however, each point in the path of the wave has been permanently displaced. The process is illustrated in Fig. 4-6 for a sawtooth wave. The

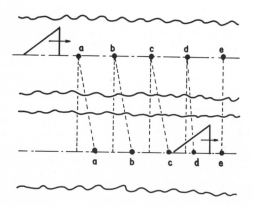

Fig. 4-6.   Displacements caused by passage of a
            sawtooth wave.

points a, b, and c over which the wave has passed
have all been displaced an equal amount, $(1/2)v_oT$,
which can be calculated by applying Eq. (4.6).
The point d which, at the instant shown, lies at the
midpoint of the advancing wave, has been displaced
an amount d_d, given by

$$d_d = \int_0^{T/2} v_o\left[1 - (1/T)t\right]dt = (3/8)v_oT.$$

Displacement as a function of time is plotted in
Fig. 4-7 a, b, and c for the square pulse, the sawtooth wave, and the flattopped wave of Fig. 4-1 a,
b, and d, respectively. In each of these cases,

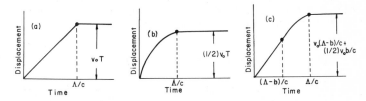

Fig. 4-7. Displacement at a point as function of
           time during passage of a transient
           stress wave: (a) square pulse; (b) saw-
           tooth wave; and (c) flattopped wave.
           Types (a), (b), and (d), respectively,
           of Fig. 4-1.

the material comes to rest after having undergone
a finite displacement. For an exponentially decaying wave, the theoretical displacement is infinite
in magnitude, which is distinctly an unrealistic
situation. Practical solutions are obtained in
such cases by adopting an arbitrary cutoff point.
    For dilatation waves, the displacement will be
in the direction of propagation of the wave, in the
same sense if the wave is one of compression and in
the opposite sense if it is one of tension. If the
transient wave has an oscillatory character, the
point will move back and forth, coming to rest at
whatever displacement solution of Eq. (4.6) yields.

MOMENTUM CONTENT

The momentum, dM, contained per unit cross section in an infinitesimal thickness of a plane longitudinal transient stress pulse will be given by

$$dM = \rho v(x)dx.$$

The total momentum M per unit cross section of the wave is then

$$M = \int_\Lambda \rho v(x)dx = \int_\Lambda (1/c)\sigma_x(x)dx. \qquad (4.7)$$

Another relation, derivable from these, is

$$\sigma_x = cdM/dx. \qquad (4.8)$$

The momentum content of waves and practical implications of momentum partitioning and transfer will be discussed much more fully in Chapter 10.

STRESS FIELD REPRESENTATION

The stresses associated with transient waves can also be depicted as a field of stresses delineated by isostress lines. Such fields are drawn (lower drawings) for a sawtooth, flattopped, and bell shaped wave in Fig. 4-8 a, b, and c, respectively. Such a form of representation is especially useful in solving problems involving the superposition of waves (Chapter 6).

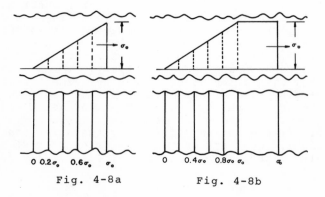

0 $0.2\sigma_0$ $0.6\sigma_0$ σ_0 0 $0.4\sigma_0$ $0.8\sigma_0$ σ_0 σ_0

Fig. 4-8a   Fig. 4-8b

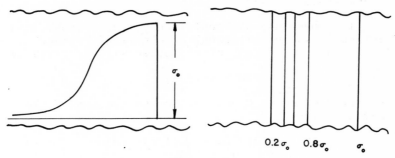

$$0.2\sigma_{_0} \quad 0.8\sigma_{_0} \qquad \sigma_{_0}$$

Fig. 4-8c

Fig. 4-8.   Fields of stress associated with (a) saw-
            tooth wave; (b) flattopped wave; and
            (c) bell shaped wave. Vertical lines in
            lower and side drawings are isostress
            lines.

PARTICLE VELOCITY FIELDS

     Every transient stress wave is a small packet
of momentum which is transported rapidly through a
material. A useful notion for describing how the
momentum is distributed at any particular instant
is that of fields of particle velocity. Using Eq.
(4.7), every point in the wave can be assigned a
particle velocity. These point by point particle
velocities taken as a group is a particle velocity
field. Two representative fields are illustrated in
Fig. 4-9, one for a sawtooth wave and the other for
a flattopped wave.

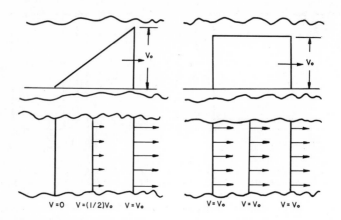

Fig. 4-9.   Fields of particle velocities associated
            with (a)   sawtooth wave; and (b) flat-
            topped wave.   Verticle lines in lower
            drawings are iso-velocity lines.

Chapter 5

SPHERICAL AND CYLINDRICAL ELASTIC WAVES

INTRODUCTION

The propagation characteristics of nonplanar
waves, such as spherically and cylindrically expanding elastic waves, are of great practical importance
because these are the types often produced by explosions and highly localized impact. Nonplanar waves
differ from plane waves in that both spherical and
cylindrical waves change shape as they advance,
altering markedly the distribution of stress and
particle velocity within the wave. A square or sawtooth compressional input pulse rapidly develops
tensile stresses and becomes oscillatory behind the
front. The fronts of the waves move with the dilatation wave velocity, c_1, with the stress, or particle
velocity, at the front of the wave decaying as $1/r$
for a spherical wave and $1/r^{\frac{1}{2}}$ for a cylindrical wave
where r is the distance from the source of the
disturbance.

In many practical problems, the front of the
wave can be treated as if it were made up of planar
elements. Over short distances, especially when at
a considerable distance from the origin of the
disturbance, the shape of the front portion of the
wave does not change enough to be of practical
importance. However, in this chapter, the wave front
is assumed curved.

Several solutions have been obtained for the
special case where a uniform time varying pressure,
p(t), is suddenly generated over the surface of a
hollow spherical or cylindrical cavity of radius a
existing in an infinite, isotropic, homogeneous
medium of density ρ and having Lamé's elastic
constants, λ and G.

37

GENERAL SOLUTION OF THE WAVE EQUATION FOR A
SPHERICAL DIVERGENT WAVE

A three dimensional wave must satisfy the wave
equation (see Eq. (3.7))

$$\partial^2 \overline{\nabla} \cdot \overline{u} / \partial t^2 = \left[(\lambda + 2G)/\rho \right] \nabla^2 \overline{\nabla} \cdot \overline{u} \qquad (5.1)$$

where $\overline{u}$ is the particle displacement vector.
For radial symmetry, the dilatation $\overline{\nabla} \cdot \overline{u}$ is given
in spherical coordinates by

$$\overline{\nabla} \cdot \overline{u} = r^{-2} \partial (r^2 u_r)/\partial r = \partial u_r / \partial r + 2u_r / r \qquad (5.2)$$

and Eq. (5.1) can be written in the form of the one
dimensional wave equation

$$\partial^2 r \overline{\nabla} \cdot \overline{u} / \partial t^2 = c_1^2 \partial^2 r \overline{\nabla} \cdot \overline{u} / \partial r^2 \qquad (5.3)$$

where

$$c_1 = \left[(\lambda + 2G)/\rho \right]^{\frac{1}{2}}$$

the velocity of dilatation waves in the medium. The
important fact that in the case of spherical waves
the quantity $r \overline{\nabla} \cdot \overline{u}$ travels with the dilatation wave
velocity without change in shape is shown here.
The principal stresses, σ_r and σ_θ, which lie in
the radial and tangenial directions, respectively,
are given by

$$\sigma_r = (\lambda + 2G) \partial u_r / \partial r + 2 \lambda u_r / r \qquad (5.4)$$

and

$$\sigma_\theta = \lambda \partial u_r / \partial r + 2 (\lambda + G) u_r / r. \qquad (5.5)$$

These equations are equivalent to Eq. (2.18) in which
σ_r and σ_θ were expressed in terms of the dilatation
$\overline{\nabla} \cdot \overline{u}$.

The mean stress, σ_m, in the solid is given by

$$\sigma_m = (1/3)(\sigma_r + \sigma_\theta + \sigma_\phi) =$$

$$= \left[\lambda + (2/3)G \right] \left[\partial u_r / \partial r + 2u_r / r \right] \qquad (5.6)$$

and the maximum shear stress, σ_s, by

$$\sigma_s = G(\partial u_r/\partial r - u_r/r) \qquad (5.7)$$

$$= (1/2)(\sigma_r - \sigma_\theta). \qquad (5.8)$$

In terms of a scalar displacement potential, ϕ, it can be shown that a spherical longitudinal pulse satisfies the wave equation

$$\partial^2\phi/\partial t^2 = c_1^2\nabla^2\phi = (1/r)\partial^2 r\phi/\partial r^2 \qquad (5.9)$$

where the displacement u_r is given by

$$u_r = \partial\phi/\partial r. \qquad (5.10)$$

For an outgoing spherical wave,

$$\phi = (A/r)f(\tau) \qquad (5.11)$$

where A is an arbitrary constant and not a function of r and t. The quantity τ is given by

$$\tau = t - (r - a)c_1. \qquad (5.12)$$

The boundary condition to be satisfied is that the radial stress σ_r at the surface of the spherical cavity, r = a, must equal the applied pressure p(t). Substitution of the relation

$$\nu = \lambda/2(\lambda + G)$$

in Eq. (5.4) puts the boundary condition in the form

$$p(t) = -\rho c^2 \left\{ [\partial u_r/\partial r] + [2\nu/(1-\nu)][u_r/r] \right\}_{r=a}. \qquad (5.13)$$

In order to simplify the solution of the problem, many workers at this point make a not especially unrealistic assumption that the Lamé's constants are equal, $\lambda = G$, which makes $\nu = 0.25$. This is not necessary, however, and several essentially equivalent general solutions are available, one of which is the following

$$\phi = (a^2/2\pi r\rho c) \int_{-\infty}^{\infty}\int \frac{p(\gamma)e^{-ix(\gamma-\tau)(c/a)}}{x^2 - ix/K - 1/K} \, dxd\gamma \qquad (5.14)$$

where

$$x = \omega a/c \qquad K = (1/2)(1 - \nu)/(1 - 2\nu)$$

and p is of the form

$$p(t) = p_o e^{i\omega t} \qquad (5.15)$$

and related to $p(\gamma)$ through the Fourier transform

$$p(t) = (1/2\pi) \int_{-\infty}^{\infty}\int p(\gamma)e^{-i\omega(\gamma-t)} \, d\gamma d\omega. \qquad (5.16)$$

SOLUTION OF SPHERICAL WAVE FOR AN EXPONENTIALLY
DECAYING PRESSURE PULSE PRODUCED BY AN EXPLOSION
IN A SPHERICAL CAVITY

The pressure pulse generated by the usual spherical explosion is a very rapid, essentially instantaneous rise to a high pressure, p_o, at the surface of the cavity which contains the explosive, followed by exponential decay. Such a pulse is expressible in the form

$$p = 0 \quad \text{for } t<0$$

$$p = p_o e^{-\alpha t} \quad \text{for } t>0 \qquad (5.17)$$

where α is a time decay constant and time is measured from the time the pressure suddenly rises. The solution for this case has the form

$$\phi = (p_o a/\bar{B}^2 \rho r)\left[-e^{-\alpha r} + \right.$$

$$\left. + (\bar{B}/\omega_o)e^{-\alpha_o r} \cos (\omega_o \tau - \beta)\right]. \qquad (5.18)$$

The various constants, α_o, ω_o, $\bar{B}^2$, and β, depend upon the material, the radius of the cavity, and the shape of the applied pressure pulse. They are given by

$$\alpha_o = (c/a)(1 - 2\nu)/(1 - \nu)$$

$$\overline{B}^2 = \omega_o^2 + (\alpha_o - \alpha)^2 \tag{5.19}$$

$$\omega_o = (c/a)(1 - 2\nu)^{\frac{1}{2}}/(1 - \nu)$$

$$\beta = \tan^{-1}[(\alpha_o - \alpha)/\omega_o]. \tag{5.20}$$

Appropriate differentiation and utilization of the relationships given in Eqs. (5.18), (5.19), and (5.20) yield expressions for radial displacement, u, radial particle velocity, v_r, radial stress, σ_r, and tangential stress, σ_θ. Such expressions can be readily arranged in a form suitable for computer calculations.

For the special case where $\lambda = G$ or $\nu = 0.25$, the potential function takes the form

$$\phi = [p_o a/\rho r(\omega/\sqrt{2} - \alpha)^2 + \omega^2]\cdot$$

$$\cdot\left\{-e^{-\alpha\tau} + e^{-\omega\tau/\sqrt{2}}\right.$$

$$\cdot\left.[(1/\sqrt{2} - \alpha/\omega)\sin \omega\tau + \cos \omega\tau]\right\} \tag{5.21}$$

where

$$\omega = 2\sqrt{2}\ c_1/3a. \tag{5.22}$$

Displacements and stresses can be obtained by appropriate differentiation of the above expression.

CONDITIONS AT THE WAVE FRONT IN A SPHERICAL WAVE

At the wave front $\tau = 0$, the following relations obtain

$$v_r = p_o a/\rho r c_1 \qquad\qquad \sigma_r = p_o a/r \tag{5.23}$$

and

$$\sigma_\theta = \lambda p_o a/(\lambda + 2G)r. \tag{5.24}$$

Thus it appears that during propagation, a discontinuous jump of stress will decrease in proportion to $1/r$ with increasing distance from the center of the cavity. This is in marked contrast with static equilibrium in which stress decreases in proportion to $1/r^3$.

Combining the two equations in Eq. (5.23) gives

$$\sigma_r = \rho c_1 v_r \qquad (5.25)$$

showing that at the wave front, radial stress and radial particle velocity are linearly related, the proportionality being the same as for a plane longitudinal wave.

Attentuation at points behind the wave front is not a simple function of r, the shape of the wave continually changing.

SPHERICAL WAVES DEVELOPED BY NONEXPONENTIALLY DECAYING PRESSURE PULSES

The response to pressure pulses having shapes differing from exponential decay such as a bell shaped wave can be obtained by application of Duhamel's theorem. The procedure is to obtain the solution for a unit step function by letting α go to zero and then applying Duhamel's integral in the form

$$U(\tau) = \frac{d}{d\tau}\int_0^\tau p(\eta)u(\tau - \eta)d\eta. \qquad (5.26)$$

The function $u(\tau)$ is the motion resulting from application of unit function pressure, $p(t) = p_0$, and $U(\tau)$ is the motion resulting from application of arbitrary pressure $p(t)$.

As α approaches zero, ϕ will be given by

$$\phi = (2ap_0/3\rho\omega^2 r)\left[-1 + \sqrt{3/2}\; e^{-\omega\tau/\sqrt{2}}\cdot\right.$$

$$\left.\cdot\sin(\omega\tau + \tan^{-1}\sqrt{2}\right]. \qquad (5.27)$$

The radial displacement, $u_r = \partial\phi/\partial r$, caused by application of the unit function pressure, is given by

$$u_r = (ap_0/4G)\left[(a/r)^2 - \sqrt{3/2}(a/r)^2\cdot\right.$$

$$\cdot e^{-\omega\tau/\sqrt{2}}\sin(\omega\tau + \tan^{-1}\sqrt{2}) +$$

$$\left. + \sqrt{2}(a/r)e^{-\omega\tau/\sqrt{2}}\sin\omega\tau\right] \qquad (5.28)$$

where as before

$$\omega = 2\sqrt{2}\ c_1/3a.$$

STRESSES ASSOCIATED WITH SPHERICAL WAVES

Calculated values of radial, tangential, and maximum shear stresses are depicted in Fig. 5-1a, b, and c as a function of t' for three values of r: r = a; r = 2a; and r/a → ∞. The units along the abcissa, given by

$$t' = ct/a - (r - a)/a \qquad (5.29)$$

have been chosen to place the graphs in register and facilitate their comparison. Since the fronts of the

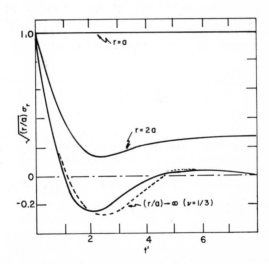

Fig. 5-1. Stresses associated with spherically expanding wave emanating from spherical cavity of radius a. Units along abcissa are t' = ct/a - (r - a)/a where c is velocity of wave front. Input pulse is of the form: σ(t) = 1, t>0; σ(t)= 0, t<0. (a) radial stress.

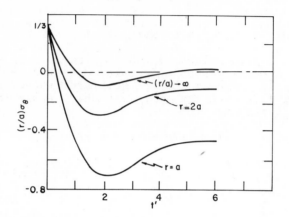

Fig. 5-1.   (b) tangential stress.

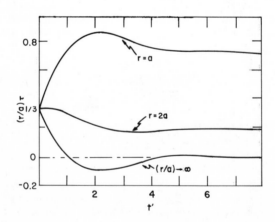

Fig. 5-1.   (c) Maximum shear stress.

waves decrease as $1/r$, convenient scales for the respective ordinates are $(r/a)\sigma_r$, $(r/a)\sigma_\theta$, and $(r/a)\sigma_s$. Poisson's ratio, ν, is taken equal to 0.25

in making the calculations, but the magnitudes and shapes of the curves are not especially sensitive to the value of ν, as is evident in Fig. 5-1a (dotted curve) where the radial stress σ_r is plotted for two values of ν, 0.25 and 0.33 for the same case of $r/a \to \infty$.

The most prominent and significant feature of the curves is that an initially positive compressive pulse must rapidly develop a negative tensile phase in order to exist in divergent propagation. This is particularly true with regard to the tangential stress which undergoes a violent stress reversal at the interface, where the tensile component in the wave becomes much larger than the compressive component.

The large shear stresses (upper curve, Fig. 5-1c) which   exist in the neighborhood of the cavity are due to the fact that in this region the negative component associated with σ_θ is considerably greater than that associated with σ_r, making the difference in the right hand term of Eq. (5.8) large.

An exponentially decaying pressure pulse produces a somewhat similar transient wave. At any given position the stresses rise at the wave front discontinuously to a peak value, then decay rapidly, oscillating about the zero value with ever decreasing amplitudes. By the third oscillation, the amplitude is essentially negligible in comparison with the maximum value attained by the variable at that point.

In both situations, $p(t) = p_o e^{-\alpha t}$ and $p = p_o$, the maximum negative tangential stress decreases from a peak value at the cavity surface with increasing distance from it. However, the radial stress in the case of the applied unit pulse does not develop a negative or tensile phase until it has traveled a distance equal to a few cavity radii, after which it becomes more and more pronounced as the wave moves out. With an applied exponential pulse, the value of the greatest tensile radial stress is zero at the cavity surface; it increases to a maximum at a distance of two or three cavity radii out, and then declines with increasing distance.

A change in decay constant α for the same cavity radius does not affect the peak positive values of the stresses.

DISPLACEMENT ASSOCIATED WITH SPHERICAL WAVES

The displacement of the surface of the cavity produced by applied pulses of different forms, all of maximum intensity p_o, is illustrated in Fig. 5-2. Each displacement was calculated using Eq. (5.28) with ν assumed equal to 0.25. The form of the applied pressure as a function of time is shown as a dotted curve and the form of the displacement of the medium as a function of time as a solid curve. The units along the abcissa are $\omega t = (2\sqrt{2}/3)(c_1/a)t$ for the pressure pulse and

$$\omega\tau = \left[2\sqrt{2}/3\right]\left[(c/a)t - (r - a)/a\right] \qquad (5.30)$$

for the displacement. The displacement ordinate is plotted in units of $[2\nu 2G/a^2 p_o]u_r$ and the pressure in units of $p(t)/p_o$.

In each case the displacement rises rapidly, but not discontinuously, to a maximum, recedes sharply, and then begins to oscillate about the zero position, meanwhile decreasing rapidly in amplitude. It is noteworthy that the shape of the displacement curve is so very insensitive to the shape of the applied pressure curve.

Examination of Eq. (5.28) shows that the displacement is made up of three terms, one of which is nonoscillatory and two of which are oscillatory. The nonoscillatory term, which represents the eventual static displacement in the medium, and the first of the oscillatory terms, depend upon the distance as $(a/r)^2$, whereas the second oscillatory term depends on distance as a/r. Consequently, the displacement at distances more than a few cavity radii is reduced simply to

$$u_r = [a^2 p_o/2\sqrt{2}Gr]e^{-\omega\tau/\sqrt{2}} \sin \omega\tau \qquad \tau > 0$$

$$u_r = 0 \qquad\qquad\qquad\qquad\qquad \tau < 0. \qquad (5.31)$$

Thus at great distances, the displacement in the wave which results from application of a unit pressure pulse of amplitude p_o is a damped sinusoidal pulse with maximum amplitude directly proportional to the pressure of the applied pulse and to the area of the cavity. The displacement will be inversely proportional to the rigidity of the medium.

The frequency f of the damped pulse is given by

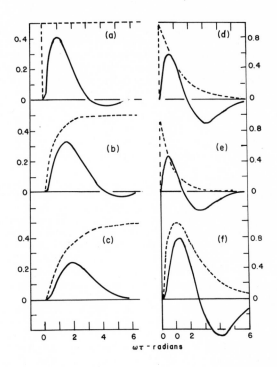

Fig. 5-2. Displacement of a point on the surface of
a cavity (solid lines) produced by application of various forms of pressure transients (dashed lines) to interior surfaces
of a spherical surface. Units along
abcissa are ωt for pressure transients and
$\omega \tau$ for displacements.
In all cases $p(t) = 0$ for $t < 0$. For $t > 0$:

(a) $p(t) = p_o$ (b) $p(t) = p_o(1-e^{-\sqrt{2}\omega t})$

(c) $p(t) = p_o(1-e^{-\omega t/\sqrt{2}})$ (d) $p(t) = p_o e^{-\omega t/\sqrt{2}}$

(e) $p(t) = p_o e^{-\sqrt{2}\omega t}$

and (f) $p(t) = N p_o(e^{-\omega t/\sqrt{2}} - e^{-\sqrt{2}\omega t})$ in which N
normalizes the maximum pressure to p_o.

$$f = \omega/2\pi = (\sqrt{2}/3\pi)(c_1/a) \qquad (5.32)$$

indicating that the frequency is directly proportional to the velocity of propagation of the wave and inversely proportional to the radius of the cavity. The damping is so high, however, that for all practical purposes the motion can be thought of as a pulse of duration Δt, given by

$$\Delta t = (3\pi/2\sqrt{2})(a/c_1). \qquad (5.33)$$

SOLUTION OF THE WAVE EQUATION FOR A CYLINDRICAL DIVERGENT WAVE

In the case of a cylindrical cavity embedded in a homogeneous, elastic, isotropic medium infinite in extent which is subjected to an impulsive internal pressure, the displacement u_r takes place in the direction of the radius vector. Applying Eqs. (5.4) and (5.5), the radial stress σ_r, the tangential stress σ_θ, and the axial stress σ_z will be given by

$$\sigma_r = (\lambda + 2G)\theta - 2Gu_r/r$$

$$\sigma_\theta = \lambda\theta + 2Gu_r/r$$

$$\sigma_z = \lambda\theta \qquad (5.35)$$

where

$$\theta = (1/r)\partial(ru_r)/\partial r. \qquad (5.36)$$

The function θ must satisfy the wave equation, giving

$$\nabla^2\theta = \partial^2\theta/\partial r^2 + (1/r)\partial\theta/\partial r = (1/c_i^2)\partial^2\theta\partial t^2. \qquad (5.37)$$

The solutions to this equation are much more complicated than those for the spherically expanding wave and will not be given in analytical form here, but the results of some specific numerical calculations are given in the next section. The wave forms obtained resemble rather closely those obtained for spherical waves.

It can also be shown that at the wave front

$$\sigma_r = (a/r)^{\frac{1}{2}}p_o \qquad (5.38)$$

and

$$\sigma_z = \sigma_\theta = \lambda/(\lambda + 2G) \, (a/r)^{\frac{1}{2}} p_o. \qquad (5.39)$$

The stresses in the wave front decrease as $r^{-\frac{1}{2}}$, in marked contrast to the r^{-2} decrease in static equilibrium.

NUMERICAL VALUES OF STRESSES ASSOCIATED WITH CYLINDRICAL WAVES

The results of a number of numerical calculations, made from solutions of Eq. (5.35) are plotted in Fig. 5-3a, b, and c. In all of these it is assumed that $\lambda = G$ or $\nu = 0.25$. The graphs show principal stresses and shear stresses, all multiplied

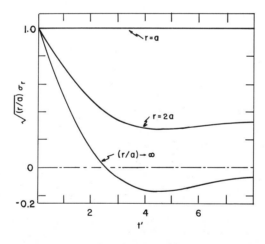

Fig. 5-3. Stresses associated with cylindrically expanding waves emanating from cylindrical cavity of radius a. Units along abcissa are t' = ct/a - (r - a)/a where c is the velocity of the wave front. Input pulse is of the form: $\sigma(t) = 1$, $t>0$ and (t) = 0, t<0.
(a) radial stress.

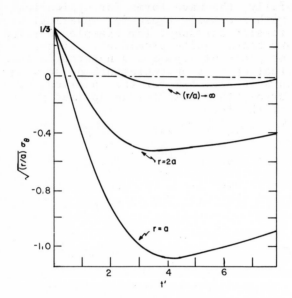

Fig. 5-3. (b) tangential stress.

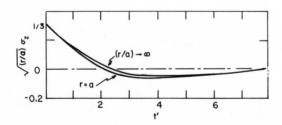

Fig. 5-3. (c) axial stress

by $(r/a)^{\frac{1}{2}}$, for different distances r from the central axis. The time parameter t' is given by

$$t' = ct/a - (r - a)/a$$

with the units of t being chosen as the time needed for the wave to travel a distance a, one unit of time being equal to $[\rho/(\lambda + 2G)]^{\frac{1}{2}}a$.

Generally, the wave forms for cylindrical waves
resemble rather closely those for spherical waves.
Negative tensile stresses, for example, develop
rapidly and become quite prominent. Compare the
stress time plots of Fig. 5-3 generated in the cylindrical situation by a unit pressure pulse with
similar ones in Fig. 5-2 for the spherical case.
Further curves, this time for an applied pressure
function of the form

$$p = e^{-0.25t} \qquad t>0$$

$$p = 0 \qquad t<0 \qquad\qquad (5.40)$$

are plotted in Fig. 5-4a, b, and c and these are
comparable to those for the spherical case given in
Fig. 5-2.

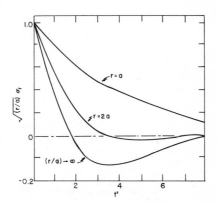

Fig. 5-4.   Similar to Fig. 5-3 except input transient
is of the form:
$$\sigma(t) = e^{-0.25t} \text{ for } t>0.$$
(a) radial stress.

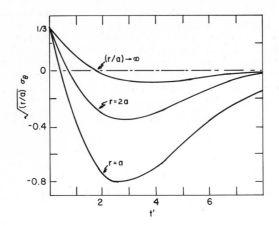

Fig. 5-4.   (b)   tangential stress.

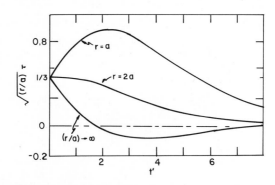

Fig. 5-4.   (c) axial stress.

The effect of the variation of the form of the input pressure on the nature of the waves generated is illustrated in Figs. 5-5 and 5-6. Radial stress at large distance, $r/a \rightarrow \infty$ is plotted in Fig. 5-5 as a function of time for exponentially decaying pressure functions of the form

$$p = e^{-Kt} \qquad t > 0$$

$$p = 0 \qquad t < 0 \qquad\qquad (5.41)$$

with the rate of decay varying from $K = 0$ to $K = 1$.
Fig. 5-6 illustrates the way in which the tangential
stress σ_θ varies at the cylinder wall, $r = a$, for the
same set of applied pressure functions. Both cases
demonstrate clearly how relatively insensitive the
stress wave patterns are to the character of the
applied pressure.

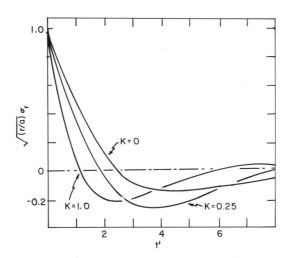

Fig. 5-5.   Similar to Fig. 5-3 with radial stress for
            $(r/a) \to \infty$ for three values of K for an
            input transient of the form:

$$\sigma(t) = e^{-Kt}.$$

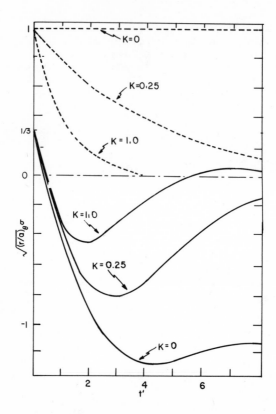

Fig. 5-6. Similar to Fig. 5-5 except tangential stress is shown for $(r/a) \to \infty$ by solid lines and for $r = a$ by dashed lines.

Chapter 6

SUPERPOSITION OF WAVES

INTRODUCTION

As long as the behavior of an elastic solid is Hookean, with stresses and strains linearly related, it is easy to describe the reaction of a body to two or more waves acting in the same region at the same time. The principle of superposition, first suggested by Lord Rayleigh, can be applied to a Hookean solid. It states that the motion resulting from the simultaneous action of any number of forces is the vector sum of the motions produced by the forces when taken separately. Superposition frequently takes place among transient stress waves and lead to the development of highly localized concentrations of stress which have important practical applications.

Interference between two or more elastic waves can arise within a body in several ways. The body initially may have been loaded simultaneously or nearly so at different locations. Or a single disturbance on encountering a boundary may be reflected or transformed leading to interference between reflected or transformed portions of the incident wave with unreflected portions. A common situation leading to interference is that in which a wave strikes two or more free surfaces, generating new disturbances at each surface which subsequently interfere.

The simplest cases to consider are those in which the directions of travel of the interfering waves are parallel; however, it is not particularly difficult to treat those cases in which the trajectories of the waves are obliquely inclined to one another.

The main cases to be considered are interference between two longitudinal waves, between a longitu-

55

dinal and a shear wave, and between two shear waves.
As long as the deformations within the body are
linear and elastic, the principle of superposition
can be used to obtain the magnitude and directions
of the stress and particle velocity at points within
the region of interference.

PARALLEL WAVES

     Consider the situation shown in Fig. 6-1 a. A
sharp fronted and sustained compressional wave of
constant stress level, σ_o, and associated particle
velocity, v_o, is assumed to be moving toward the
left with velocity c_1, and a tensile wave of the
same absolute magnitude of stress is assumed to be
moving with the same velocity toward it from the
opposite direction. Shortly after the parallel two
waves have met (Fig. 6-1 b), tension and compression

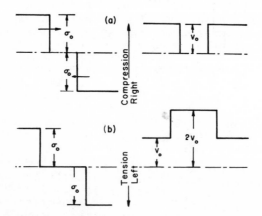

Fig. 6-1.   Interference between a sharp fronted
            compression wave and a sharp fronted
            tension wave of constant stress level
            σ_o and associated particle velocity v_o.
            (a) just prior to encounter and (b)
            shortly after encounter.

will annul each other in the region of interaction,
reducing the stress to zero in that region. Particle

motion is directed the same in both waves, to the
left, thus within the region of overlap, the
material will be moving to the left with velocity
$2v_o$.
    When two like waves, either a pair of parallel
compressional waves or a pair of parallel tensile
waves of equal and sustained stress level meet
(Fig. 6-2), the stress will double in the region of
overlap and the particle velocity will be zero.

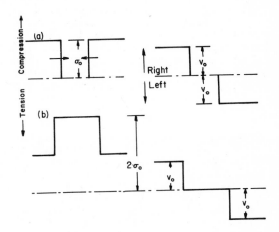

Fig. 6-2.   Similar to Fig. 6-1 except both waves
            are compressive.

    Disturbances of finite duration and variable
stress can be treated in essentially the same way.
In the region of interaction or overlap, the precise
spatial distribution of stress and particle velocity
will depend upon the shape of the interacting stress
waves. As an example, consider the interaction
between two identical sawtooth compression waves
traveling toward each other (Fig. 6-3). At the
instant the wave fronts meet, the stress along the
plane where they meet immediately doubles. As the
fronts continue to move, the value of the stress in
the region of overlap steadily decreases, finally
reaching zero when the waves pass beyond one another.

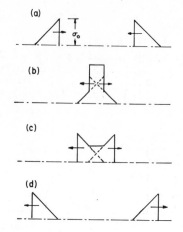

Fig. 6-3.   Similar to Fig. 6-1 except interference
            is between two compressive sawtooth
            waves. Particle velocity distributions
            are not shown. (a), (b), (c), and (d)
            represent successive times.

The principle of superposition can also be
applied to interference between two parallel transverse waves but the problem is complicated by the
fact that the motion in transient shear waves is
generally polarized in a particular direction in the
plane of the wave front, the directions in the two
wave fronts usually being different. Simple
algebraic addition is then not possible and other
methods for combining stresses must be used, such as
are described in the following section.

OBLIQUELY INTERSECTING WAVES

Generally, two plane wave fronts intersect
obliquely and this situation is best treated using
Mohr circle diagrams for combining stresses. The two
waves may be either a pair of dilatation waves, a
pair of shear waves, or one of each. The problem is
to develop equations for calculating the magnitudes
and orientations of the principal stresses, σ, σ',
and σ'', generated by each type of interaction.

The principal stresses associated with a single dilatation wave are

$$\sigma_i \,, [\nu/(1 - \nu)]\sigma_i \,, [\nu/(1 -\nu)]\sigma_i$$

where σ_i is the stress lying normal to the wave front. For a shear wave of stress level τ_i, the principal stresses will be σ_i, $-\sigma_i$, and 0 where $\sigma_i = \tau_i$. These principal stresses will be oriented at an angle of 45° with the wave front. The sign conventions used are that the normal stresses are assumed positive in compression and that the shearing stresses are positive if, on the edges of the element away from the coordinate axes, they are directed in the positive coordinate direction.

First consider the case where two plane dilatation waves, with respective stress levels of σ_1 and σ_2 interact. The problem is to determine the principal stresses, σ, σ', and σ'', and their orientations due to the superposition of two states of stress, the principal stresses of which are, respectively

$$\sigma_1 \,, \sigma_1\nu/(1 - \nu) \,, \sigma_1\nu/(1 - \nu)$$

and

$$\sigma_2 \,, \sigma_2\nu/(1 - \nu) \,, \sigma_2\nu/(1 - \nu).$$

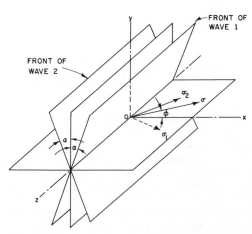

Fig. 6-4.   Choice of reference planes for solution of superposition of two stress waves.

For convenience, the three axes are chosen as shown
in Fig. 6-4. The axis oz lies along the line formed
by the intersection of the two planes that are,
respectively, normal to the directions of propagation of the two waves; the axes ox and oy lie in the
two planes that bisect these planes.

From symmetry, one of the principal stresses
must be parallel to oz. Its value, σ'', is given by

$$\sigma'' = (\sigma_1 + \sigma_2)\nu/(1 - \nu). \tag{6.1}$$

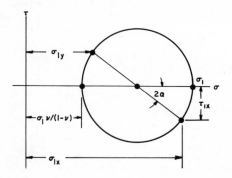

Fig. 6-5.   Mohr's circle for state of stress
            due to wave 1.

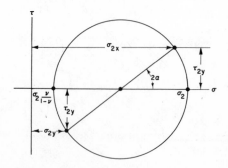

Fig. 6-6.   Mohr's circle for state of stress
            due to wave 2.

The other two principal stresses, σ and σ' are in the xy plane and can be determined using Mohr's circle. This is done by first projecting σ_1 and σ_2 onto the xz and yz planes which results in tensile and shear components, σ_{1x}, σ_{1y}, τ_{1x}, and τ_{1y}, corresponding to σ_1; and σ_{2x}, σ_{2y}, τ_{2x}, and τ_{2y}, corresponding to σ_2 (Figs. 6-5 and 6-6). The shear stresses, τ_{1x} and τ_{2x}, lie in a plane perpendicular to oy. Addition of stress components for the plane perpendicular to ox gives for the stresses σ_x and τ_x in this plane

$$\sigma_x = \sigma_{1x} + \sigma_{2x}$$

$$\tau_x = \tau_{1x} + \tau_{2x} \qquad (6.2a)$$

and for σ_y and τ_y, the stresses associated with the plane perpendicular to oy,

$$\sigma_y = \sigma_{1y} + \sigma_{zy}$$

$$\tau_y = \tau_{1y} + \tau_{zy}. \qquad (6.2b)$$

Since σ_1 and σ_2 are known and the orientation of the axes ox, oy, and oz have been fixed with respect to the directions of propagation of the waves, the stresses σ_x, τ_x, σ_y, and τ_y can be calculated and from these the principal stresses σ and σ' determined. Mohr's circle is applied three times: once to determine σ_{1x}, τ_{1x}, σ_{1y}, and τ_{1y} from σ_1 ( ); a second time to determine σ_{2x}, τ_{2x}, σ_{2y}, and τ_{2y} from σ_2 ( ); and finally to determine σ and σ' from σ_x, τ_x, σ_y, and τ_y ( ).

An examination of Fig. 6-7 shows that

$$\sigma_{1x} = [\sigma_1 + \sigma_1 \nu/(1 - \nu)]/2 +$$

$$+ \left\{ [\sigma_1 - \sigma_1 \nu/ (1 - \nu)]/2 \right\} \cos 2\alpha \qquad (6.3)$$

$$\tau_{1x} = -\left\{ [\sigma_1 - \sigma_1 \nu/(1 - \nu)]/2 \right\} \sin 2\alpha \qquad (6.4)$$

with comparable expressions for σ_{1y} and τ_{1y}, all of which reduce to

$$\sigma_{1x} = [\sigma_1/2(1 - \nu)][1 + (1 - 2\nu) \cos 2\alpha] \qquad (6.5)$$

$$\sigma_{1y} = [\sigma_1/2(1 - \nu)][1 - (1 - 2\nu) \cos 2\alpha] \qquad (6.6)$$

$$\tau_{1x} = -[\sigma_1(1 - 2\nu)/2(1 - \nu)] \sin 2\alpha \qquad (6.7)$$

$$\tau_{1y} = [\sigma_1(1 - 2\nu)/2(1 - \nu)] \sin 2\alpha \qquad (6.8)$$

where α is equal to one half the angle between the two wave fronts.

Similarly, from Fig. 6-6, it follows that for the second wave:

$$\sigma_{2x} = [\sigma_2/2(1 - \nu)][1 + (1 - 2\nu) \cos 2\alpha] \qquad (6.9)$$

$$\sigma_{2y} = [\sigma_2/2(1 - \nu)][1 - (1 - 2\nu) \cos 2\alpha] \qquad (6.10)$$

$$\tau_{2x} = [\sigma_2(1 - 2\nu)/2(1 - \nu)] \sin 2\alpha \qquad (6.11)$$

$$\tau_{2y} = -[\sigma_2(1 - 2\nu)/2(1 - \nu)] \sin 2\alpha. \qquad (6.12)$$

Substitution of these expressions into Eqs. (6.2a) and (6.2b) yields

$$\sigma_x = [(\sigma_1 + \sigma_2)/2(1-\nu)][1 + (1-2\nu) \cos 2\alpha] \qquad (6.13)$$

$$\sigma_y = [(\sigma_1 + \sigma_2)/2(1-\nu)][1 - (1-2\nu) \cos 2\alpha] \qquad (6.14)$$

$$\tau_x = [(\sigma_2 - \sigma_1)(1 - 2\nu)/2(1 - \nu)] \sin 2\alpha \qquad (6.15)$$

$$\tau_y = [(\sigma_1 - \sigma_2)(1 - 2\nu)/2(1 - \nu)] \sin 2\alpha. \qquad (6.16)$$

From Fig. 6-7, a Mohr's circle plot of σ_x, σ_y, τ_x, and τ_y, it is evident that one principal stress σ is given by

$$\sigma = (\sigma_x + \sigma_y)/2 + [(\sigma_x - \sigma_y)^2/4 + \tau_x^2]^{\frac{1}{2}} \qquad (6.17)$$

$$= (\sigma_1 + \sigma_2)/2(1 - \nu) +$$

$$+ [(\sigma_1 + \sigma_2)^2(1-2\nu)^2 \cos^2 2\alpha/4(1-\nu)^2 +$$

$$+ (\sigma_2 - \sigma_1)^2(1-2\nu)^2 \sin^2 2\alpha/4(1-\nu)^2]^{\frac{1}{2}} \qquad (6.18)$$

which reduces to

$$\sigma = \frac{\sigma_1 + \sigma_2 + (1-2\nu)[\sigma_1{}^2 + \sigma_2{}^2 + 2\sigma_1\sigma_2 \cos 4\alpha]^{\frac{1}{2}}}{2(1 - \nu)} \qquad . \quad (6.19)$$

Similarly, the other principal stress σ' becomes

$$\sigma' = \frac{\sigma_1 + \sigma_2 - (1-2\nu)\left[\sigma_1{}^2 + \sigma_2{}^2 + 2\sigma_1\sigma_2 \cos 4\alpha\right]^{\frac{1}{2}}}{2(1 - \nu)} \qquad . (6.20)$$

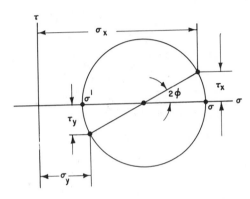

Fig. 6-7.   Mohr's circle for state of stress due to the superposition of waves 1 and 2.

The angle ϕ between ox and one of the principal stresses is determined by

$$\tan 2\phi = 2\tau_x/(\sigma_x - \sigma_y) \qquad (6.21)$$

obtained by examination of Fig. 6-7. On substitution of expressions for τ_x, σ_x, and σ_y in terms of σ_1 and σ_2 this becomes

$$\tan 2\phi = \left\{(\sigma_2 - \sigma_1)/(\sigma_2 + \sigma_1)\right\} \tan 2\alpha. \quad (6.22)$$

Figure 6-8 is a reference drawing showing the geometrical relationships existing among wave patterns and stresses.

For the particular case where the two intersecting stress waves are equal in magnitude, $\sigma_1 = \sigma_2$, the principal stresses of the resulting state of stress will be

$$\sigma = [\sigma_1/(1 - \nu)][1 + (1 - 2\nu)\cos 2\alpha] \qquad (6.23)$$

$$\sigma' = [\sigma_1/(1 - \nu)][1 - (1 - 2\nu)\cos 2\alpha] \qquad (6.24)$$

$$\sigma'' = 2\sigma_1\nu/(1 - \nu). \qquad (6.25)$$

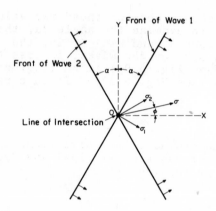

Fig. 6-8.   Reference drawing showing relation
            between angle ϕ, ox axis, and direction
            of principal stress σ.

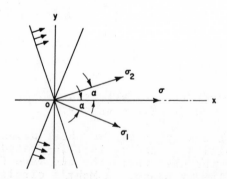

Fig. 6-9.   Direction of highest principal stress when
            angle 2α between directions of propagation
            of the two stress waves is less than 90°.

The directions of σ and σ', determined by

$$\tan 2\phi = \left[(\sigma_2 - \sigma_1)/(\sigma_2 + \sigma_1)\right]\tan 2\alpha = 0 \qquad (6.26)$$

become

$$\phi = 0 \text{ or } 90^\circ.$$

In order to choose between these two values of ϕ, it is necessary to examine the angle 2α, the angle included between σ_1 and σ_2. If 2α, the angle of intersection between the waves, is less than 90°, ϕ equals 0 (see Fig. 6-9); if 2α is greater than 90°, ϕ equals 90° (see Fig. 6-10). If 2α equals 90° then

$$\tan 2\phi = 0 \times \infty.$$

and ϕ is undetermined; all directions in the resulting state of stress are principal and hydrostatic compression or tension exists.

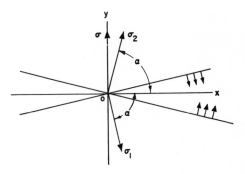

Fig. 6-10. Similar to Fig. 6-9 except for angles 2α greater than 90°.

The intersection of two shear waves of respective intensities τ_1 and τ_2 is illustrated in Fig. 6-11 a for the case where the two wave fronts are normal to the xy plane. A Mohr's circle analysis similar to that used for solving the dilatation-dilatation interaction can be applied to obtain the magnitudes and orientation of the resulting princi-

pal stresses. The appropriate Mohr's circles are
drawn in Fig. 6-11 b and c. Combining them as before
gives for the principal stresses

$$\sigma = \tau_1 - \tau_2 \qquad\qquad (6.27)$$

$$\sigma' = \tau_2 - \tau_1. \qquad\qquad (6.28)$$

The angle ϕ that one of the principal stresses makes
with the ox axis is simply

$$\phi = 90 - \alpha.$$

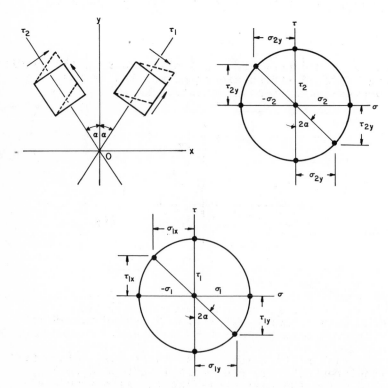

Fig. 6-11. Mohr's circle construction for state of
            stress due to superposition of two
            shear waves.

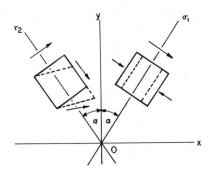

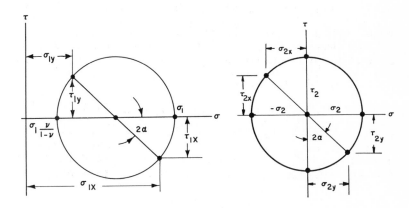

Fig. 6-12.   Mohr's circle construction for state of
             stress due to superposition of a shear
             wave and a dilatation wave.

   The intersection of a dilatation wave of stress
level σ_1 and a shear wave of stress level τ_2 is
illustrated in' Fig. 6-12 a with appropriate Mohr's
circles given in Fig. 6-12 b and c. The solution for
the principal stresses is

$$\sigma = (1/2)\left\{\sigma_1/(1-\nu) + \left[\sigma_1{}^2(1-2\nu)^2/(1-\nu)^2 + \right.\right.$$
$$\left.\left. + 4\sigma_1\tau_1(1-2\nu)\sin 4\alpha/(1-\nu) + 4\tau_2{}^2\right]^{\frac{1}{2}}\right\} \qquad (6.29)$$

$$\sigma' = (1/2)\left\{\sigma_1/(1-\nu) - \left[\sigma_1{}^2(1-2\nu)^2/(1-\nu)^2 + \right.\right.$$
$$\left.\left. + 4\sigma_1\tau_1(1-2\nu)\sin 4\alpha/(1-\nu) + 4\tau_2{}^2\right]^{\frac{1}{2}}\right\} \qquad (6.30)$$

The angle ϕ between ox and one of the principal stresses can be determined from the relationship

$$\tan 2\phi = \frac{\sigma_1(1 - 2\nu)/2(1 - \nu) \tan 2\alpha + \tau_2}{\sigma_1(1 - 2\nu)/2(1 - \nu) + \tau_2 \tan 2\alpha}. \qquad (6.31)$$

STRESS FIELDS DEVELOPED BY SUPERPOSITION

It was pointed out in Chapter 4 that it is often useful to represent the stress field of a plane transient stress wave by a set of parallel and constant stress level planes. Such sets of planes are convenient in delineating the temporary stress field established by the interference of two transient stress waves. To define such a field requires knowledge of the magnitudes and direction of the principal stresses developed at every point within the region of interference.

Figure 6-13 illustrates how a calculation might proceed for two interfering sawtooth waves of respective maximum stress levels σ_1 and σ_2 and relative wave front inclination 2α. A set of constant stress level planes is drawn for each wave. The grid can be as fine as desired but in the illustration six steps are used. The construction defines an area

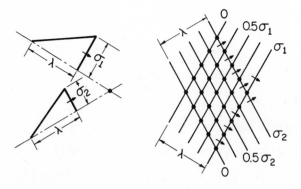

Fig. 6-13.   Pattern of superposition of two sawtooth stress transients.

containing thirty-six points of intersection. Using
Eqs. (6.19) and 6.20), the magnitudes and directions
of the principal stresses can be calculated for each
point. From this group of calculations, it is then a
straightforward matter to draw isostress and iso-
clinic lines.

Chapter 7

Effects of Boundaries on Elastic Wave Propagation

INTRODUCTION

The waves discussed thus far have been assumed to be traveling in bodies of infinite extent. Most bodies have boundaries. Sooner or later, the wave will encounter one or more of them and will suffer transformations as a result of the interactions.

The transformations, reflections, and refractions occur at boundaries because of changes in physical properties occurring across them. The two main mechanical properties involved are the elastic and inertial characteristics of the materials in which the waves are traveling. The explicit parameters that are needed to describe quantitative aspects of the transformations are the densities and wave velocities, the latter dependent both upon the elastic constants and the density. The product of density and wave velocity, $\rho_i c_i$, plays an important role and has been called the specific acoustic resistance. The values of the product ρc listed in Table 4-1 for a few different materials are useful in making numerical calculations.

Although it is possible to approach the problem of wave interactions with boundaries by working out the general solution and from this deriving equations for special situations, the course followed here is to develop equations from scratch for specific situations since this approach seems more instructive. Five different situations are looked at: normal incidence at a free boundary; normal incidence at a boundary between dissimilar materials; oblique incidence at a free boundary; oblique incidence at a cohesive boundary between dissimilar materials; and oblique incidence at a noncohesive boundary.

70

NORMAL INCIDENCE AT FREE BOUNDARY

When a plane wave strikes a free boundary
normally, it is totally reflected with a 180° change
in phase. Thus a compression wave will be reflected
as a tension wave and vice versa. A free surface is
incapable of supporting normal and shear stresses
and must therefore remain stress free during the
interaction of the wave with the free surface. The
surface, being unconfined on one side, will acquire
a particle velocity equal to twice the particle
velocity of the interacting wave.

Generally the wave will be of finite duration.
As it is reflected, the head portion of the reflected
wave will at first superpose itself on the tail
portion of the incident wave. Finally it emerges as

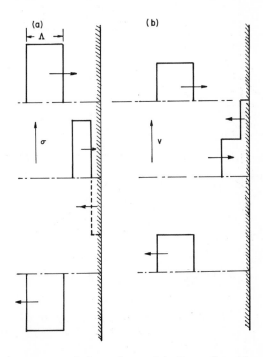

Fig. 7-1.   Stress (a) and particle velocity (b)
distributions developed during normal
reflection of a square transient pulse
at a free boundary.

a complete wave moving in a direction opposite to
the direction of motion of the incident wave. The
case of a square compressive pulse of length Λ is
illustrated in Fig. 7-1 and that of a sawtooth pulse
in Figs. 7-2 and 7-3. For the sawtooth wave, the
distribution of stress is shown in Fig. 7-4 as a
function of time where it is plotted at four selected
points: (a) a distance Λ from the boundary; (b) Λ/2
from the boundary; (c) Λ/4 from the boundary; and
(d) on the boundary. Note that tension builds up as
the wave moves to the left.

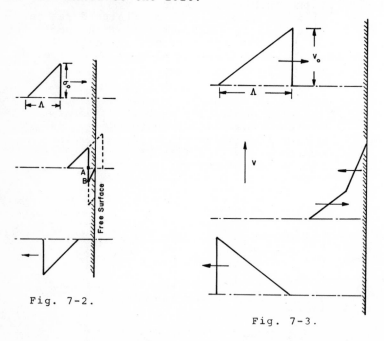

Fig. 7-2.

Fig. 7-3.

Fig. 7-2.   Stress distribution developed during
             normal reflection of a sawtooth wave
             at a free boundary.

Fig. 7-3.   Particle velocity distribution developed
             during normal reflection of sawtooth
             wave at free boundary.

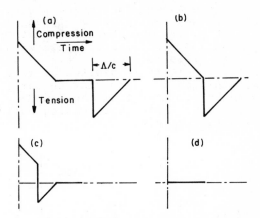

Fig. 7-4.   Stress as a function of time at four
            selected points during reflection of a
            sawtooth wave at a free boundary: (a)
            Λ from the boundary; (b) Λ/2 from the
            boundary; (c) Λ/4 from the boundary; and
            (d) at the boundary.

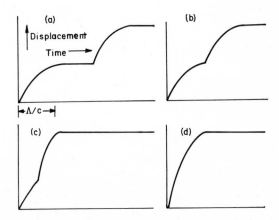

Fig. 7-5.   Similar to Fig. 7-4 except displacements
            of the four respective fronts are shown
            as function of time.

Displacement as a function of time is shown in Fig. 7-5 for the same points. The final displacement of each point is the same but for points within a distance of $\Lambda/2$ of the surface, the paths by which the points reached their respective final displacements are quite different.

The stresses, particle velocities, and displacements generated during the reflection of a transient pulse of any arbitrary shape can be found graphically in essentially the same way as has been illustrated for square and sawtooth waves.

NORMAL INCIDENCE AT BOUNDARY BETWEEN DISSIMILAR MATERIALS

When a plane elastic wave strikes a plane interface between two dissimilar materials, the interaction is regulated by two boundary conditions. First, the stresses on the two sides of the boundary must be equal at every instant during the interaction; and second, the normal particle velocities on both sides of the boundary must be equal. The first condition results from the fundamental law of hydrostatic pressures and it can be shown that as long as normal incidence alone is considered, it must also hold for solids. The second condition is equivalent to saying that the two media remain in constant contact at the boundary.

The two equations that express the above conditions can be written as

$$\sigma_I(x,t) + \sigma_R(x,t) = \sigma_T(x,t) \qquad (7.1)$$

$$v_I(x,t) + v_R(x,t) = v_T(x,t) \qquad (7.2)$$

where σ_I, σ_R, σ_T, v_I, v_R, and v_T are the instantaneous values of stress and particle velocity, respectively, for the incident, reflected, and transmitted waves, respectively.

From Eq. (4.4) it follows that

$$v_I = \sigma_I/\rho c_1 \ , \ v_R = -\sigma_R/\rho c_1 \ , \ v_T = \sigma_T/\rho' c_1' \qquad (7.3)$$

where ρ and c are the density of the material and velocity of propagation of the wave, respectively, and the primes denote the second medium.

Substituting Eq. (7.3) into Eq. (7.2) yields

$$\sigma_I/\rho c_1 - \sigma_R/\rho c_1 = \sigma_T/\rho'c_1'. \qquad (7.4)$$

Solving Eq. (7.1) and Eq. (7.4) simultaneously, first for σ_T in terms of σ_I and then for σ_R in terms of σ_I, two fundamental equations governing the partitioning of stress at an abrupt change in media are obtained:

$$\sigma_T = [2\rho'c_1'/(\rho'c_1' + \rho c_1)]\sigma_I \qquad (7.5)$$

$$\sigma_R = [(\rho'c_1' - \rho c_1)/(\rho'c_1' + \rho c_1)]\sigma_I \qquad (7.6)$$

From these, the ratio of transmitted to reflected stress is given by

$$\sigma_T/\sigma_R = 2\rho'c_1'/(\rho'c_1' - \rho c_1). \qquad (7.7)$$

A number of conclusions can be drawn from Eqs. (7.5) and (7.6). When the two products ρc_1 and $\rho'c_1'$ are equal, the ratio σ_R/σ_I is zero and there is no reflected wave. The incident wave is transmitted at full intensity such as when the material is identical on both sides of the boundary. When $\rho c_1 < \rho'c_1'$, the ratio σ_R/σ_I is positive, implying that if σ_I is originally a compressive wave, the reflected wave will also be compressive. When $\rho c_1 > \rho'c_1'$, compressive waves will be reflected as tension waves and vice versa, provided the joint can support tension. When $\rho'c_1'$ is zero, the condition for a free surface, $\sigma_R = -\sigma_I$, a compressive wave is reflected at full stress level as a tension wave and vice versa. The transmitted stress (Eq. (7.5)) will always have the same sign as the incident stress, compression resulting in compression and tension in tension, provided, of course, that the joint can support tension. If it cannot, a tensile stress will be reflected at the boundary as a compressive stress, the second medium acting as if it were not there at all. When the second body is completely rigid, $\rho'c' = \infty$, the stress felt by the rigid body will be just twice that of the incident stress, the reflected stress being equal to the incident stress.

The variation of the ratio of transmitted to incident stress is plotted in Fig. 7-6 as a function of the ratio of the respective specific acoustic resistances of the two materials. Values were

calculated from Eq. (7.7). For ratios of specific
acoustic resistance of less than one, the transmitted stress is less than the incident stress. For
ratios of specific acoustic resistance greater than
one, the ratio of transmitted to incident stress is
greater than one, reaching asymtotically a value of
two. Several numerical examples of stress partit-
tioning in specific cases are given in Table 7-1
which lists the nature and magnitude of the several
stresses for various combinations of common materials.

Table 7-1. Partition of stress for a plane
            compression wave incident on a bi-
            metallic interface (normal incidence).

From (medium 1)	Into (medium 2)	σ_R	σ_T
		(in terms of σ_I)	
Steel	Aluminum	-0.46	+0.54
	Brass	-0.12	+0.88
	Lead	-0.31	+0.69
	Magnesium	-0.61	+0.39
Aluminum	Brass	+0.36	+1.36
	Lead	+0.17	+1.17
	Magnesium	-0.21	+0.79
	Steel	+0.46	+1.46
Brass	Aluminum	-0.36	+0.64
	Lead	-0.19	+0.81
	Magnesium	-0.53	+0.47
	Steel	+0.12	+1.12
Lead	Aluminum	-0.17	+0.83
	Brass	+0.19	+1.19
	Magnesium	-0.37	+0.63
	Steel	+0.31	+1.31
Magnesium	Aluminum	+0.21	+1.21
	Brass	+0.53	+1.53
	Lead	+0.37	+1.37
	Steel	+0.61	+1.61

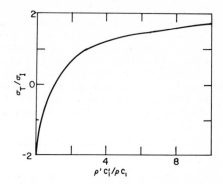

Fig. 7-6. Ratio of transmitted to incident stress as
           function of ratio of specific acoustic
           resistance for normal incidence.

The reflection and transmission of a sawtooth
wave striking an interface between two similar
materials is illustrated in Figs. 7-7 and 7-8 for
two different situations. In the first (Fig. 7-7),
the specific acoustic resistance of the material in
which the wave originates is greater than that of the
material into which the wave enters; in the second

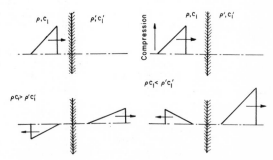

Fig. 7-7. Stress partitioning occurring as a
           consequence of interaction of a sawtooth
           wave with a boundary between dissimilar
           materials where $\rho c_1 > \rho' c_1'$.
Fig. 7-8. Similar to Fig. 7-7 except $\rho c_1 < \rho' c_1'$.

(Fig. 7-8), the respective relative values of the
specific acoustic resistances are reversed. In both
cases, the wave divides its momentum, part being
transmitted and part being reflected, the extent of
partitioning depending upon the relative values of
the specific acoustic resistance in accordance with
Eqs. (7.5) and (7.6). The lengths of the disturbance in the two media will usually be different
since these depend upon the velocity of propagation
of the wave. In the first case, the compression
wave is reflected at the boundary as a tension wave
because of the lower specific acoustic resistance of
the second material, whereas in the second case,
again a compressive wave, it is reflected as a
compression wave since the specific acoustic resistance of the second material is greater than that
of the first.

More intricate interactions will occur when
several materials of different specific acoustic
resistances are juxtaposed, as in layered structures.
These interactions can be treated in substantially
the same way: the distribution of stress is obtained geometrically using the principle of superposition and Eqs. (7.5) and (7.6) for computing the
relative magnitudes of the several transmitted and
reflected stresses. It can be shown through such
considerations that layers thin with respect to the
length of the wave,Λ, will have substantially no
effect on the wave. Thus a wave will be transmitted
nearly as effectively through an assembly of like
blocks with thin layers of cement between them as
through a solid single block of the same material,
the cement in the interfaces having essentially no
effect on the wave. Likewise, the cement layers
between dissimilar blocks will not affect the wave
interactions taking place at the joints.

OBLIQUE INCIDENCE AT FREE BOUNDARY

Oblique reflection of a plane longitudinal wave
from a free plane surface is illustrated in Fig. 7-9.
The boundary conditions are that the normal and shear
stresses on the surface must be zero at all times
although the tangential stress need not necessarily
be zero. Both a longitudinal and a shear wave
develop. The angle at which the newly generated
longitudinal wave goes off from the surface is equal

to the angle with which the incident wave strikes
the surface, but the shear wave comes off at a
different angle, given by

$$\sin \beta = (c_2/c_1) \sin \alpha \qquad (7.8)$$

a result following directly from Snell's law. This
law, most often met with in optics, describes in
mathematical form the direction the wave front
assumes as it moves through a body in which the wave
velocity changes. The angle β is the angle between
the normal to the front of the shear wave and the
normal to the free surface, and α is the angle between the normal to the front of the incident longitudinal wave and the normal to the free surface.
The shear wave is polarized in the plane of
incidence.

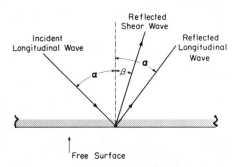

Fig. 7-9.   Ray geometry for oblique reflection of a
            dilatation wave from a free surface.
            Reflected dilatation wave and reflected
            shear wave are shown.

During the transformation of the original
longitudinal wave into a longitudinal and shear wave,
the momentum of the original wave is partitioned
between the two waves with the respective amounts of
energy in each wave depending upon both angle of
incidence and Poisson's ratio of the material. The
partitioning is most conveniently expressed quantitatively by means of a reflection coefficient R.
Through application of the boundary conditions, it
can be shown that the following relationships must
obtain:

$$\sigma_R + R\sigma_I \qquad (7.9)$$

$$\tau_R = \left[(R + 1) \cot 2\beta\right]\sigma_I \qquad (7.10)$$

where σ_R is the magnitude of the reflected longitudinal stress and τ_R the magnitude of the reflected shear stress. The coefficient R is given by

$$R = \frac{\tan \beta \cdot \tan^2 2\beta - \tan \alpha}{\tan \beta \cdot \tan^2 2\beta + \tan \alpha} \ . \qquad (7.11)$$

Curves giving the coefficient of reflection versus angle of incidence for various values of Poisson's ratio are drawn in Fig. 7-10. A negative R indicates a reversal in longitudinal stress, a compressive wave becoming a tensile wave and vice versa. For the limiting case of normal incidence, $\alpha = 0$, reflection is complete and no shear wave is

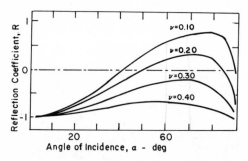

Fig. 7-10.   Reflection coefficient R as function of angle of incidence α for various Poisson's ratios, ν, for oblique reflection of a dilatation wave at a free surface.

generated. While it might appear that the same situation obtains for the other limiting case, $\alpha = 90°$, it does not. This matter is discussed later. Note again that the relative intensity of the longitudinal and shear waves is strongly dependent upon angle of incidence and Poisson's ratio, so much so that under certain conditions, for example, $\alpha = 50°$ and $\nu = 0.20$, all of the energy of the longitudinal

wave is transformed into shear wave energy. In fact,
the development of the reflected shear wave may have
so much influence that the reflected longitudinal
stress does not change in sign at the boundary, a
compression wave remaining a compression wave, as is
the case for $\alpha = 70^{\circ}$ and $\nu = 0.20$.
    A point on the free surface acted upon by the
incident wave will move in a direction $\bar{\alpha}$, given by

$$\bar{\alpha} = \tan^{-1} (v_x/v_y) \qquad (7.12)$$

where v_x is the vector sum of the particle velocities
of the three waves in the plane of the surface, and
v_y, the sum normal to the surface. The angle $\bar{\alpha}$ is
known as the angle of emergence and is equal to 2β.
In Fig. 7-11, the angle of emergence is plotted
against angle of incidence for several Poisson's
ratios. The principal thing to notice is that
particle motion in general is not normal to the free
surface, rather it lies nearly but not exactly normal to the incident wave front. The obliquity of the
particle velocity produces tangential stress along
the free surface, a condition favorable to the generation of a surface wave.

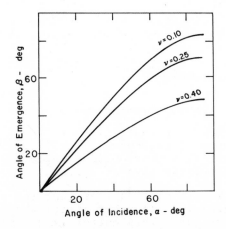

Fig. 7-11.   Angle of emergence, β, as a function of
             angle of incidence α for various
             Poisson's ratios, ν.

Two things can happen when a transverse wave strikes obliquely. If the wave is polarized with particle motion in the incident wave perpendicular to the plane of incidence, no longitudinal wave is formed and the transverse wave is reflected unaltered. If, however, some particle motion occurs in the plane of incidence, a longitudinal wave will also be generated and the inclination of its wave front can be calculated from Eq. (7.8). For large angles of incidence, total reflection occurs, the critical angle β_c at which this starts being given by

$$\beta_c = \sin^{-1} (c_2/c_1). \qquad (7.13)$$

For most materials, β_c is about 30°.

If particle motion in the incident wave is restricted to the plane of incidence, the intensity τ_R of the reflected shear wave will be given by

$$\tau_R = \tau_I R \qquad (7.14)$$

and the intensity σ_R of the reflected longitudinal wave by

$$\sigma_R = [(R - 1) \tan 2\beta]\tau_I \qquad (7.15)$$

where τ_I is the intensity of the incident shear wave and R is the same reflection coefficient as before, with β representing the angle of incidence in this case. In Fig. 7-12, R is plotted as a function of

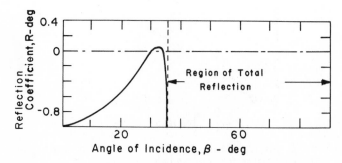

Fig. 7-12.   Reflection coefficient R for oblique reflection of a shear wave at a free surface. Angle of incidence, β.

β for ν = 0.25 between the limits of 0<β<35°, the
region where total reflection does not take place.
    In case the wave is polarized in a plane lying
between the plane of incidence and a plane perpendicular to it, the respective motions can be
resolved vectorially.

OBLIQUE INCIDENCE AT COHESIVE BOUNDARY BETWEEN
TWO DISSIMILAR MATERIALS

    When a longitudinal wave arrives at a cohesive
boundary between two dissimilar materials, part of
the energy will be transmitted and part will be
reflected. At the point of encounter, O in Fig. 7-13
usually four new waves, C, D, E, and F are created

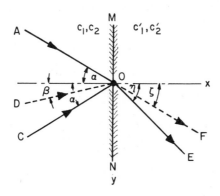

Fig. 7-13. Ray geometry for reflection and trans-
            mission of obliquely incident dilatation
            wave, A, at cohesive interface, MN,
            between two dissimilar materials. C is
            reflected dilatation wave; D, reflected
            shear wave; E, transmitted dilatation
            wave; and F, transmitted shear wave.

between the original wave A and the boundary MN. The
relationships among these waves are illustrated in
the same figure. Two of these are longitudinal
waves, one returning into the original material and
the other crossing the boundary to be transmitted
through the second material. Similarly two transverse waves are generated.

The angles at which these waves move away from the boundary are in accordance with Snell's law given by the relations

$$\sin \alpha/c_I = \sin \beta/c_2 = \sin \eta/c_1' = \sin \zeta/c_2'$$

where α, β, η, and ζ are the respective angles of incidence for the five waves involved, A and C, D, E, and F, c_1 and c_1' are the respective velocities of the longitudinal waves in the medium to the left of the boundary MN in Fig. 7-13 and to the right of the boundary, and c_2 and c_2' are the respective transverse wave velocities in the two media.

If the interface is slipless, the boundary conditions that must be met are continuity of normal displacement, continuity of tangential displacement, continuity of normal stress, and continuity of tangential stress across the interface. Application of these boundary conditions results in the following set of equations:

$$A\cos\alpha - C\cos\alpha + D\sin\beta - E\cos\eta - F\sin\zeta = 0 \qquad (7.17)$$

$$A\sin\alpha + C\sin\alpha + D\cos\beta - E\sin\eta + F\cos\zeta = 0 \qquad (7.18)$$

$$A\cos 2\beta + C\cos 2\beta - D(c_2/c_1)\sin 2\beta -$$

$$- E(\rho'/\rho)(c_1'/c_1)\cos 2\zeta -$$

$$- F(\rho'/\rho)(c_2'/c_1)\sin 2\zeta = 0. \qquad (7.19)$$

$$A\sin 2\alpha - C\sin 2\alpha - D(c_1/c_2)\cos 2\beta -$$

$$- E(\rho'/\rho)(c_2'/c_2)^2(c_1/c_1')\sin 2\eta +$$

$$+ F(\rho'/\rho)(c_2'/c_2)^2(c_1/c_2')\cos \zeta = 0 \qquad (7.20)$$

where A is the amplitude of the incident longitudinal wave; C, of the reflected longitudinal wave; D, of the reflected shear wave; E, of the transmitted longitudinal wave; and F, of the transmitted shear wave. These four equations can be solved for the amplitudes C, D, E, and F of the newly generated waves in terms of the amplitude A of the incident wave. The solutions, in determinant form, are

$$C/A = \frac{\begin{vmatrix} -\cos\alpha & \sin\beta & -\cos\eta & -\sin\zeta \\ -\sin\alpha & \cos\beta & -\sin\eta & \cos\zeta \\ \cos 2\beta & (c_2/c_1)\sin 2\beta & (\rho'c_1'/\rho c_1)\cos 2\zeta & (\rho'c_2'/\rho c_1)\sin 2\zeta \\ \sin 2\alpha & (c_1/c_2)\cos 2\beta & (\rho'/\rho)(c_2'/c_2)^2(c_1/c_1')\sin 2\eta & -(\rho'/\rho)(c_2'/c_2)^2(c_1/c_2')\cos 2\zeta \end{vmatrix}}{|K|}$$

(7.21)

$$D/A = \frac{\begin{vmatrix} -\cos\alpha & -\cos\alpha & -\cos\eta & -\sin\zeta \\ \sin\alpha & -\sin\alpha & -\sin\eta & \cos\zeta \\ -\cos 2\beta & \cos 2\beta & (\rho'c_2'/\rho c_1)\cos 2\zeta & (\rho'c_2'/\rho c_1)\sin 2\zeta \\ \sin 2\alpha & \sin 2\alpha & (\rho'/\rho)(c_2'/c_2)^2(c_1/c_1')\sin 2\eta & -(\rho'/\rho)(c_2'/c_2)^2(c_1/c_2')\cos 2\zeta \end{vmatrix}}{|K|}$$

(7.22)

$$E/A = \frac{\begin{vmatrix} -\cos\alpha & \sin\beta & -\cos\alpha & -\sin\zeta \\ \sin\alpha & \cos\beta & -\sin\alpha & \cos\zeta \\ -\cos 2\beta & (c_2/c_1)\sin 2\beta & \cos 2\beta & (\rho'/\rho)(c_2'/c_1)\sin 2\zeta \\ \sin 2\alpha & (c_1/c_2)\cos 2\beta & \sin 2\alpha & (\rho'/\rho)(c_2'/c_2)^2(c_1/c_2')\cos 2\zeta \end{vmatrix}}{|K|}$$

(7.23)

$$F/A = \frac{\begin{vmatrix} -\cos\alpha & \sin\beta & -\cos\eta & -\cos\alpha \\ \sin\alpha & \cos\beta & -\sin\eta & -\sin\alpha \\ -\cos 2\beta & (c_2/c_1)\sin 2\beta & (\rho'c_1'/\rho c_1)\cos 2\zeta & \cos 2\beta \\ \sin 2\alpha & (c_1/c_2)\sin 2\beta & (\rho'/\rho)(c_2'/c_2)^2(c_1/c_1')\sin 2\eta & \sin 2\alpha \end{vmatrix}}{|K|}$$

(7.24)

$$|K| = \begin{vmatrix} -\cos\alpha & \sin\beta & -\cos\eta & -\sin\zeta \\ \sin\alpha & \cos\beta & -\sin\eta & \cos\zeta \\ -\cos 2\beta & (c_2/c_1)\sin 2\beta & (\rho'c_1'/\rho c_1)\sin 2\zeta & (\rho'c_2'/\rho c_1)\sin 2\zeta \\ \sin 2\alpha & (c_1/c_2)\cos 2\beta & (\rho'/\rho)(c_2'/c_2)^2(c_1/c_1')\sin 2\eta & -(\rho'/\rho)(c_2'/c_2)^2(c_1/c_2')\cos 2\zeta \end{vmatrix}$$

(7.25)

The determinant solutions have not been expanded into algebraic expressions because computations are more easily made directly from the determinants.

A shear wave striking a boundary obliquely, illustrated in Fig. 7-14, will also be transformed with new shear and longitudinal or dilatation waves developing. Since any transient shear wave can be resolved into two components, one of which is parallel to the plane of incidence, xy in the figure, and the other which is perpendicular to it, parallel to the z axis, it is sufficient to consider only these two cases. In seismology these two components or types of shear waves are designated SV- and SH- waves, respectively.

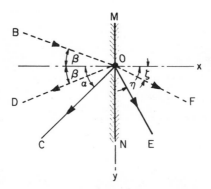

Fig. 7-14.   Ray geometry for reflection and transmission of obliquely incident shear wave, B, at cohesive interface, MN, between two dissimilar materials. C is reflected dilatation wave; D, reflected shear wave; E, transmitted dilatation wave; and F, transmitted shear wave.

The partitioning of amplitudes in the case of a shear wave in which particle motion is parallel to the plane of incidence can be obtained by solving the following set of equations for the ratios C/B, D/E, E/B, and F/B:

$$-B\sin\beta - C\cos\alpha + D\sin\beta - E\cos\eta - F\sin\zeta = 0 \qquad (7.26)$$

$$B\cos\beta + C\sin\alpha + D\cos\beta - E\sin\eta + F\cos\zeta = 0 \qquad (7.27)$$

$$B \sin 2\beta - C(c_1/c_2) \cos 2\beta + D \sin 2\beta +$$

$$+ E(\rho'c_1'/\rho c_2) \cos 2\zeta +$$

$$+ F(\rho'c_2'/\rho c_2) \sin 2\zeta = 0 \qquad (7.28)$$

$$-B \cos 2\beta + C(c_2/c_1) \sin 2\alpha + D \cos 2\beta +$$

$$+ E(\rho'c'^2_2/\rho c_2 c_1') \sin 2\eta -$$

$$- F(\rho'c_2'/\rho c_2) \cos 2\zeta = 0. \qquad (7.29)$$

The four boundary conditions applied to obtain the above equations are the same as those applied for the case of a dilatation wave.

When particle motion in the shear wave is perpendicular to the plane of incidence, the displacements and stresses normal to the interface vanish. The only two boundary conditions left are continuity of particle velocity in the z direction along the interface and continuity of displacement in the z direction. Application of these boundary conditions lead to the following equations governing the partitioning of amplitudes:

$$B + D = F \qquad (7.30)$$

$$B - D = [\rho'c_2 \cos \zeta/\rho c_2 \cos \beta]F. \qquad (7.31)$$

No dilatation waves are generated.

It is apparent from Eq. (7.16) that for large angles of incidence, $\sin \eta$ will become greater than 1 whenever c_1' is greater than c_1, and that both $\sin \eta$ and $\sin \zeta$ will be greater than 1 when c_2' is greater than c_1. This is a physically impossible situation which results in total reflection. As soon as the critical angle is reached at which the normal to the wave front, often referred to as a ray, grazes the interface, that ray ceases to exist as a refracted ray.

OBLIQUE INCIDENCE AT NONCOHESIVE BOUNDARY

When a boundary between two blocks of material is frictionless and can slip freely, the boundary conditions to be applied are naturally different than for a slipless boundary. Stress can be transmitted only normal to the boundary. This will be true for blocks of either similar or dissimilar material. Since, by definition, a non cohesive boundary can support no tension, an obliquely incident tension wave will be reflected as if the boundary were a free surface. For an obliquely incident compressive wave, the following boundary conditions must be met: continuity of normal displacement on the two sides of the interface, continuity of normal stress across the interface, and absence of shear stresses in the media on both sides of the interface. In general, four additional waves are generated during the interaction of the incident wave with the boundary.

Applying these boundary conditions when the material on both sides of the loose boundary are similar, yields the following relations among wave amplitude ratios:

$$C/A = \frac{\sin 2\alpha \sin 2\beta}{(K^2 \cos^2 2\beta + \sin 2\alpha \sin 2\beta)} \qquad (7.32)$$

$$D/A = F/A = \frac{K \cos 2\beta \sin 2\alpha}{(K^2 \cos^2 2\beta + \sin 2\alpha \sin 2\beta)} \qquad (7.33)$$

$$E/A = \frac{K^2 \cos^2 2\beta}{(K^2 \cos^2 2\beta + \sin 2\alpha \sin 2\beta)} \qquad (7.34)$$

where

$$K = c_1/c_2 \qquad (7.35)$$

and A, C, D, E, and F are, respectively, the amplitudes of the incident compressive dilatation wave, the reflected dilatation wave, the reflected shear wave, the transmitted dilatation wave, and the transmitted shear wave.

The ratios of particle velocities of the respective waves are the same as the ratios between amplitudes. Thus,

$$v_C/v_A = C/A \qquad (7.36)$$

$$v_D/v_A = v_F/v_A = D/A = F/A \qquad (7.37)$$

$$v_E/v_A = E/A. \qquad (7.38)$$

Some typical relative particle velocity, or relative amplitude, curves are plotted for a low Poisson's ratio, 0.25, in Fig. 7-15, and for a high Poisson's ratio, 0.40, in Fig. 7-16. At high angles of obliquity and a low Poisson's ratio, proportionately more of the momentum of the original wave ends up in the reflected and transmitted shear waves (which are equal) than in the case of a material having a high Poisson's ratio.

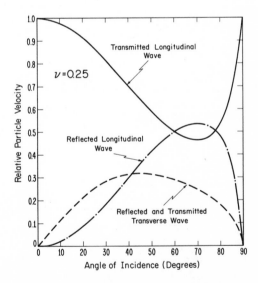

Fig. 7-15. Partitioning of particle velocity among several waves involved in reflection of obliquely incident dilatation wave at loose joint between two blocks of same material.

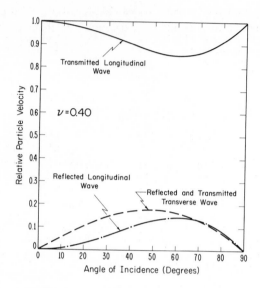

Fig. 7-16.   Similar to Fig. 7-15 except for higher
              Poisson's ratio, 0.40.

Generally the two sides of a loose joint will
move with respect to each other, such as ab and cd
in Fig. 7-17. Using values of particle velocity
calculated from the above equations or taken off
curves similar to those of Figs. 7-15 and 7-16, it
is a straightforward matter to calculate the motion

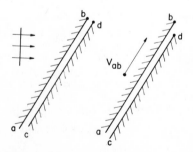

Fig. 7-17.   Motion along loose joint caused by
              obliquely incident dilatation wave.
              v_{ab} is velocity of face ab with respect
              to face cd.

that takes place. The pattern of waves and associated particle velocity components developed during
the encounter of the wave with the joint is illustrated in Fig. 7-18.

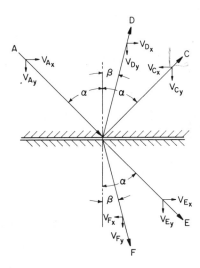

Fig. 7-18.   Ray geometry and particle velocity
             components associated with dilatation
             wave striking loose joint obliquely.
             A, C, D, E, and F correspond to the
             same type of waves as in Fig. 7-14.

     The right hand face (cd in Fig. 7-17) of the
joint will move only perpendicular to itself. Its
velocity will be the vector sum of v_E and v_F (Fig.
7-18). The face ab of Fig. 7-17 will also have the
same component of velocity in the same direction but
in addition will have a component of velocity parallel to the interface. This parallel component v_{ab}
will be given by the expression

$$v_{ab} = v_A \sin \alpha + v_D \cos \beta - v_C \sin \alpha. \qquad (7.39)$$

The right hand side of this expression is the sum of
the particle velocity components along the interface
of each of the three waves involved in the interaction which reside in the upper side of the joint.
The velocity v_{ab} has been plotted in Fig. 7-19 as a

as a function of angle of incidence for several
Poisson's ratios.

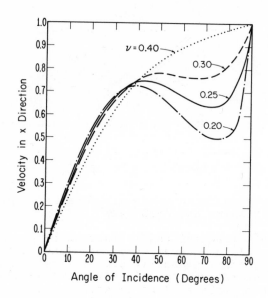

Fig. 7-19.   Velocity of upper face of joint of
             Fig. 7-18 (v_{ab} of Fig. 7-17) as function
             of angle of incidence for several
             Poisson's ratios. The unit one on the
             ordinate corresponds to particle velocity
             in incident wave.

     For a transient compression wave, the total
displacement d of the face ab with respect to the
face cd (Fig. 7-17) will be given by

$$d = \int v_{ab}(t)dt. \qquad (7.40)$$

where the integration is taken over the duration of
the wave. The relative displacement of surfaces
that would be caused by a square topped transient
wave of unit particle velocity and unit duration
impinging at a 45° angle is shown in Fig. 7-20. The
Poisson's ratio is assumed to be 0.25.
     A common effect of an underground nuclear
explosion is the development at the time of the
explosion of displacements along jointed rock

structures and preexisting faults. Calculations
using the above integrals give quantitative results
that agree well with field observations.

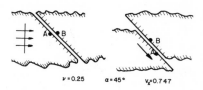

$\nu = 0.25$ $\alpha = 45°$ $V_x = 0.747$

Fig. 7-20.   Displacement, A with respect to B,
             caused by square topped transient of
             unit particle velocity and unit duration
             impinging at   45° angle.

CURVED FRONTED WAVES AT PLANE AND CURVED
BOUNDARIES AND INTERFACES

The reflection and refraction of non planar
waves at plane boundaries or of plane waves at non
planar boundaries usually can be treated in only a
semi quantitative way. An effective stratagem is
to think of each surface or wave front as made up
of an infinite number of infinitesimally small
planar elements, each element tangent to the surface
at a point. The problem then resolves itself into
describing a set of interactions with each interaction being between only two planar elements,
usually non parallel.
Since these interactions are between planar
elements, all of the foregoing equations in this
chapter can be used to calculate the partitioning
of energy among the new dilatation and shear waves
generated at each point. Also, since for most
curved wave fronts and curved surfaces the orientation between planar elements will vary from point to
point, the relative amounts of energy going into
each type of wave will change from point to point.
Often in making drawings of the interactions of
waves with boundaries, the thicknesses of the lines
depicting the wave fronts are varied to give a
qualitative indication of different intensities of
stress along the wave fronts. This matter will be
discussed in more detail in Chapter 11.

REFLECTION OF PLATE WAVES

A dilatation wave incident on a free surface producing reflected dilatation and shear waves, and the partitioning of energy has been described above. Similar partitioning occurs for a plane pulse traveling in a thin plate and reflected from a free edge. In this latter case, the pulse length will be considered to be large compared with the thickness of the plate to make the problem tractable.

It is assumed in deriving the equations that govern the partitioning that the pulse is traveling in an infinitely wide plate which is normal to the z axis, that the edge of the plate (Fig. 7-21) lies

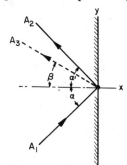

Fig. 7-21.   Ray geometry for reflection of plate wave at free boundary. A_1 is incident dilatation wave, A_2 is reflected dilatation wave, and A_3 is reflected shear wave.

in the yz plane, that the displacements in the x and y directions are independent of z, and that the stress normal to the plate, σ_z, is everywhere zero, a condition of plane stress. The boundary conditions which are $\sigma_x = \tau_{xy} = 0$ at $x = 0$ will be satisfied if

$$\sin \alpha_1/c_1 = \sin \alpha_2/c_1 = \sin \beta/c \qquad (7.41)$$

giving by inspection, $\alpha_1 = \alpha_2 = \alpha$. The boundary conditions, $\sigma_x = 0$ for all t and y, yields

$$(A_1 + A_2) \cos 2\beta \sin \alpha - A_3 \sin 2\beta \sin \beta = 0 \qquad (7.42)$$

where A_1, A_2, and A_3 are the respective amplitudes of
the three pulses, incident dilatation, reflected
dilatation, and reflected shear. The other boundary
condition, $\tau_{xy} = 0$ at $x = 0$, gives

$$2(A_1 - A_2) \sin \beta \cos \alpha - A_3 \cos 2\beta = 0. \qquad (7.43)$$

The ratios of amplitudes $-(A_2/A_1)$ and (A_3/A_1)
are plotted for the particular case of $c_1/c_2 = 1.76$
in Fig. 7-22. The ratio A_3/A_1 increases rapidly
with increasing obliquity, reaching a maximum at
about 45º while the ratio A_2/A_1 decreases rapidly
over the same range. Thus in the neighborhood of
45º, practically all of the incident dilatation wave
energy is transformed into shear wave energy.

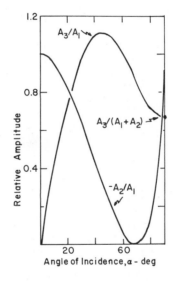

Fig. 7-22.   Relative amplitudes of waves of
             Fig. 7-21 as function of angle of
             incidence, α.

CHAPTER 8

DEVELOPMENT OF SURFACE WAVES, PLATE WAVES, AND BAR WAVES

INTRODUCTION

By combining Newton's laws of motion with the elastic relations obtained from Hooke's law, it can be shown that in an unbounded solid two different types of elastic waves can be propagated, a dilatation wave and a distortion wave (Chapter 4). Dilatation waves involve changes in volume but no rotation. Distortion waves, on the other hand, do not cause any changes in volume but result in both rotation and shearing of the medium and travel with a lower velocity, usually about one half that of the dilatation wave velocity. The particle motion in the case of a plane dilatation wave is along the direction of propagation and the wave is often referred to as a longitudinal wave; the motion in a distortion wave is at right angles to the direction of propagation and the wave is termed transverse or shear.

In order to derive the equations that govern the propagation of waves in unbounded solids it is necessary to introduce the appropriate boundary conditions. In most cases this leads to equations which are mathematically intractable. In some simple cases exact solutions have been found and the propagation of a longitudinal wave along a cylinder is one of these. Even for the simplest case, the computation is complex and it is only recently that calculations have been carried out. They show that for the fundamental mode of vibration the phase velocity of plane sinusoidal waves along a cylinder is given by $c_o = (E/\rho)^{\frac{1}{2}}$ for waves of infinitely long wavelength, where E is Young's modulus and ρ is density, and by

the velocity of Rayleigh surface waves, c_s, for
waves of infinitely short wavelength. Waves of
finite wavelength travel at velocities between these
two limits. In general, the propagation of an
elastic disturbance down a cylinder will differ from
propagation in an infinite medium only insofar as the
waves are being continually reflected at the sides
of the cylinder. In between reflections they must
travel with either the dilatation wave velocity, c_1,
or the shear wave velocity, c_2.

This chapter shows, largely from geometrical
considerations, how surface waves, plate waves, and
bar waves evolve out of the reflections and interactions of dilatation and shear waves.

WAVE MOVING ALONG BOUNDARY

Experimentally it is observed that a sharp
fronted plane longitudinal wave moving along a free
surface at glancing incidence generates a strong
shear wave, generally referred to as a head wave.
The reflection coefficient curves of Fig. 7-10 do not
predict the existence of such a wave and, in essence,
its generation is part of diffraction phenomena
controlled by the presence of a free boundary rather
than true reflection phenomena. While a quantitative mathematical treatment is not feasible, the
pattern of wavelets and waves that might be expected
to develop under these conditions is fairly easy to
understand in a qualitative way from simple geometrical considerations.

The following discussion of waves along boundaries is based on Fig. 8-1 which depicts a plane
compressive wave originating along the plane MN and
then moving to the right at glancing incidence along
the free surface MK of an elastic body. Elements
lying in the path of the wave will be compressed as
the wave front AB passes over them. Those elements
lying at the surface are free to expand. Whereas
an interior particle, such as C in the same figure,
initially acquires a velocity lying wholly in the
direction of motion of the wave front, the initial
movement of a particle on the surface A will be both
upward and forward. The value of Poisson's ratio
will regulate the relative amplitudes of the hori-
zontal and vertical motions.

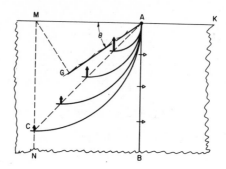

Fig. 8-1.   Mechanics of generation of trailing shear
             wave AG and expansion waves bounded by AC
             caused by movement of a dilatation wave
             front AB along a free surface.
             Angle θ is equal to $\sin^{-1}(c_2/c_1)$.

The upward expansion spawns two wavelets as it
passes at each point, such as A, one a dilatation
disturbance propagating at velocity c_1, and the other
a shear disturbance propagating at velocity c_2.
Application of Hyygen's principle fixes the positions
and configurations of the new wave fronts.
    The dilatation wavelets do not form a true
coherent front, for it is not possible to construct
a Huygen's envelope for them. Qualitatively, the
region in which expansion has been felt can be
thought of as a region containing an infinite number
of wavelets each having originated at a new point
along the surface as the parent wave progresses along
it. While each wavelet could be presumed to have no
energy, the combined effect of all the wavelets is to
extract energy from the parent wave and to introduce
a region, bounded on the left by the circular arc AC
in the figure, behind the wave front in which lateral
motion as well as forward motion is present. Material lying beteen AC and the wave front will be moving strictly normal to the wave front. The particle
motion transmitted by each wavelet will in general
be maximum in a direction perpendicular to the free
surface. There should develop therefore a plane,
AC, inclined at 45° to the surface, which is the
locus of maximum lateral particle velocities.

Maximum lateral particle velocity will decrease
rapidly with increasing distance from the surface,
quickly becoming negligible in value.

The shear wave generated by the expansion of
the surface is relatively strong. The individual
shear wavelets combine to form a coherent wave front
AG and particle motion at the surface is controlled
by the respective velocities, c_1 and c_2, of the
dilatation and shear waves being given by

$$\sin \theta = c_2/c_1 = (1 - 2\nu)/2(1 - \nu)^{\frac{1}{2}}. \quad (8.1)$$

Hence θ is a function of Poisson's ratio alone, a
physically reasonable conclusion. The amplitude of
the shear wave will be essentially zero to the left
of the plane that lies normal to the shear wave
front and that passes through the intersection of the
free surface and the plane MN in which the parent
wave originated.

Both the shear wave and the expansion wavelets
derive their energies from the parent compressive
wave; thus the intensity of the wave front is being
continually degraded. Eventually the portion of the
front near the surface will disappear altogether.
The final outcome of the interaction, the details of
which are very complex and will not be traced, is
the generation of a Rayleigh surface wave, described
below, which travels with a velocity slightly less
than the shear wave velocity.

RAYLEIGH SURFACE WAVES

Transient disturbances, moving along the free
surface of an elastic half space, leave in their
wakes surface waves that are similar to gravitation-
al surface waves in liquids. The waves, now known
as Rayleigh waves, were first investigated by Lord
Rayleigh in 1887. He showed that the amplitude of
the wave decreased rapidly with increasing distance
from the surface and that the velocity of propagation, c_s, will be given by

$$c_s = \kappa_1 c_2 \quad (8.2)$$

where κ_1^2 must satisfy the cubic equation

$$\kappa_1^6 - 8\kappa_1^4 + (24 - 16\alpha_1^2)\kappa_1^2 + (16\alpha_1^2 - 16) = 0 \qquad (8.3)$$

and α_1 is a function only of Poisson's ratio, given by

$$\alpha_1 = (1 - 2\nu)/(2 - 2\nu). \qquad (8.4)$$

The velocity of propagation is independent of frequency and depends only on the elastic constants of the material. There is no dispersion of these waves and a plane surface wave will travel without change in form.

The value of κ_1 in all cases is slightly less than one so that in any particular material the Rayleigh wave travels somewhat slower than the distortion wave. For Poisson's ratios equal to 0.25, 0.29, and the extreme case 0.50, κ_1 assumes, respectively, the values of 0.9194, 0.9258, and 0.9554. For particles at the surface, the ratio between the major and minor axes of the ellipse is 1.468 ($\nu = 0.25$) and the motion is retrograde, a particle tracing out its elliptic path in a counterclockwise sense as the wave moves from left to right. The major axis is normal to the surface.

The amplitude of the displacement in the direction of propagation decreases with increasing distance from the surface, the rate of attenuation depending on frequency. It becomes zero at a depth of 0.193 wavelengths ($\nu = 0.25$), and for greater depths becomes finite again though of the opposite sign so that the vibrations then take place in the opposite phase.

Motion in a direction normal to the surface also varies with distance from the surface. As the distance increases, the amplitude of the vibration first increases, reaching a maximum at a depth of 0.076 wavelengths ($\nu = 0.25$) and then decreases monotonically. At a distance of one wavelength, the amplitude has fallen to 0.19 of its value at the surface. Rayleigh waves of high frequency will attenuate more rapidly with depth than those of low frequency.

Calculated values of the amplitude of displacements for steel, whose Poisson's ratio is 0.29, are plotted in Fig. 8-2 as a function of distance from the surface. The curves are given in nondimensional form, the peak value u of the displacement in the direction of propagation of the wave and the peak value w of the displacement perpendicular to the

direction of propagation of the wave being plotted
as the ratios u/w_o and w/w_o where w_o is the amplitude of the vibrations perpendicular to the surface
at the surface. These ratios are plotted against
z/Λ where Λ is the wavelength of the vibrations.
The curves illustrate the nature of the attenuation
with the frequency variations mentioned above.

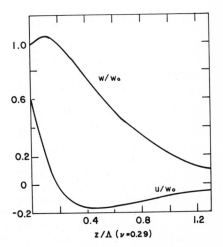

Fig. 8-2.   Relative displacement in a Rayleigh wave
            as a function of relative depth, z/Λ, for
            steel. u is displacement in direction of
            propagation of wave and w is displacement
            normal to direction of propagation. w_o
            is amplitude of perpendicular displace-
            ment at surface, z is depth, and Λ is
            wavelength.

     Rayleigh waves play a prominent role in the
analysis of earthquake records. They spread only in
two dimensions and their intensity falls off much
less rapidly than dilatation and distortion waves.

WAVE MOVING THROUGH A PLATE

     A step pressure pulse applied to the edge of a
plate of finite thickness and infinite lateral extent
starts a plane fronted dilatation wave of compression

through the plate. The parent wave spawns tensile
and shear waves at the two boundary surfaces which
begin to interact in complex ways, rapidly degrading
the original front of the wave and setting up
oscillations. The abrupt front is soon transformed
into a sloping front whose rise time is commensurate
with the transit time of a wave across the plate.

After the wave has traveled a considerable
distance (20 or so plate thicknesses), the bulk of
the energy is moving through the plate with a
velocity c_p, the so-called plate velocity, which is
given by

$$c_p = c_2[2/(1-\nu)]^{\frac{1}{2}} = c_1[(1-2\nu)/(1-\nu)^2]^{\frac{1}{2}}. \quad (8.5)$$

For most metals, Poisson's ratios are about 0.3, so
that in general, c_p is roughly 20 percent less than
the dilatation wave velocity.

The frequency of the oscillations which are
superimposed on the main transport of energy are
regulated by the thickness of the plate and its
elastic properties. These oscillations are clearly
visible as waves on the surface of a plate subjected
to an impulsive load,under special lighting conditions.

Applying geometric constructions simultaneously
to both sides of the plate, it is apparent that a
pattern of internal waves will develop appearing
roughly as drawn in Fig. 8-3. The parent compressive
dilatation wave front AB trails two shear waves of
which BD and AC are segments. These trailing shear
waves are reflected at the free surfaces without
change in intensity since they arrive at the critical
angle of incidence. The reflected segments such as
DE and CF begin new oblique traverses across the
plate.

At any particular instant, the distribution of
stress and particle velocity is very inhomogeneous
but it possesses a periodicity regulated by the
thickness of the plate and the velocity with which
the elastic wave propagates. The distances AD, DF,
BC, and CE which are all equal are related to plate
thickness, D, wave velocities, and Poisson's ratio
by

$$\overline{AD} = D \cot \theta = D[(c_1/c_2)^2 - 1]^{\frac{1}{2}} \quad (8.6)$$

$$= D(1 - 2\nu)^{-\frac{1}{2}}. \quad (8.7)$$

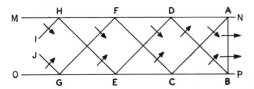

Fig. 8-3.   Mechanics of development of internal waves
             in plate or rod.   AB is dilatation wave
             front.   IH, JG, GF, etc., are shear wave
             fronts originating at free surface MN and
             OP.   Arrows indicate direction of motion
             of respective wave fronts.

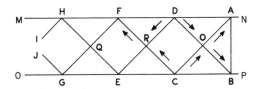

Fig. 8-4.   Arrows indicate directions of particle
             velocities associated with waves depicted
             in Fig. 8-3.

The particle velocities associated with the
several segments of shear waves are shown in Fig.8-4.
Here the element DOCR is being compressed in a
direction parallel to the line DC whereas the element
FREQ is being extended in the same direction.   The
result is that as the wave propagates to the right,
each point on the plate oscillates up and down with
a frequency, f, where

$$f = c_1/\overline{HD} = c_1(1 - 2\nu)^{\frac{1}{2}}/D. \qquad (8.8)$$

The variation of frequency with plate thickness,
HD, is plotted in Fig. 8-5.   For a 5 mm thick glass
plate, with a Poisson's ratio of 0.25 and a dilatation wave velocity of 5400 m/sec, the frequency is
756,000 hertz.   A particle velocity or displacement
sensor placed on the surface of the plate would
record repetitive changes occurring with the above
frequency.

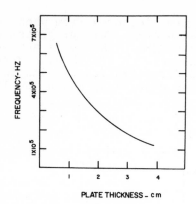

Fig. 8-5.   Plot of Eq. (8.8), frequency of plate
            wave as function of plate thickness for
            $v = 0.25$ and $c_1 = 5400$ m/sec.

    As the wave front AB in Fig. 8-4 moves forward,
the region affected by the shear wave lengthens, with
more and more segments looking like DOCR being generated.   At first the front AB moves with the
dilatation wave velocity but it soon, within two to
three plate thicknesses, slows down to plate velocity
as the effects of lateral motion are felt.   Viewed
qualitatively, lateral motion reduces resilience by
lessening the effects of rigidity so that the wave
is traveling in a softer medium and hence more slowly.   Thus the wave pattern of Fig. 8-3 is not
strictly accurate.   A more accurate picture would
show the rearward shear wave segments inclined somewhat more steeply since the ratio between the shear
wave velocity and the plate wave velocity is greater
by about 20 percent than the ratio between the shear
wave velocity and the dilatation wave velocity.   This
tends to increase slightly the frequency of the
oscillatory disturbance.
    As the wave progresses through the plate, the
separation between the front AB and the last or tail
of the shear wave segments, points I and J, increases.
The velocity with which the tail moves forward is
less than the velocity of the front AB of the pulse.
The separation L between head and tail at time t is

$$L = (c_1 - c_2 \sin \theta)t \qquad (8.9)$$

where t is the time measured from when the wave
first enters the plate. This progressive separation
is plotted in Fig. 8-6. The total number of principal oscillations which a sensor placed on a particular point would detect increases with increasing
distance from the edge along which the load is
applied.

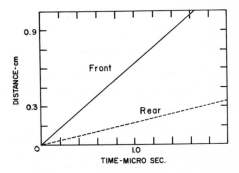

Fig. 8-6. Progressive separation of front of pulse
           (AB in Fig. 8-3), solid line; and rear of
           pulse (IJ in Fig. 8-3), dashed line.
           $\nu = 0.25$ and $c_1 = 5400$ m/sec.

The situation to the left of the tail (I and J
in Figs. 8-3 and 8-4) is difficult to assess even
qualitatively. It is only possible to treat it as a
diffraction problem. The conditions are ideal for
the excitation of vibrations through the thickness of
the plate.

GLANCING ANGLE REFLECTION OF ELASTIC WAVES
FROM A FREE BOUNDARY

The treatment of the reflection of a dilatation
wave from a free boundary, given in Chapter 7, results in a seemingly trivial solution for glancing
incidence, $\alpha = 90°$, where all motion and stress
vanish. Experimentally, it is found that a pulse
advancing at glancing incidence generates a strong,

trailing shear wave, often referred to as the von
Schmidt head wave. It is also apparent from the
Huygen's construction of Fig. 8-1 that the generation of the shear wave is a diffraction phenomenon
controlled by the presence of the free boundary.
    While the reflection method cannot be applied
near the onset of the boundary, it can be applied
further along when, for all practical purposes
the front of the expansion wave has merged with the
original wave front and the region AOB of Fig. 8-4
ceases to exist. A solution can be obtained by
using a limiting process to determine the ratios
$A_3/(A_1 + A_2)$ and $A_3/(A_1 - A_2)$ where A_1 is the
amplitude of the incident dilatation wave, A_2 is the
amplitude of the reflected dilatation wave, and A_3
is the amplitude of the newly generated shear wave.
The amplitudes refer to either particle displacement
or particle velocity, these two quantities being
proportional. The ratios will be given by

$$A_3/(A_1 + A_2) = \sin \alpha_1 \cos 2\beta_2/\sin \beta_2 \sin 2\beta_2 \qquad (8.10)$$

and

$$A_3/(A_1 - A_2) = 2\cos \alpha_1 \sin \beta_2/\cos 2\beta_2 \qquad (8.11)$$

where α_1 is the angle of incidence of the wave and
β_2 is the angle of emergence of the shear wave.
This simple treatment of reflection results in a
solution in which the flow of energy from the pressure pulse to the shear pulse is constant and finite.
    The value which the angle β_2 attains for glanc-
ing incidence depends upon whether it is a two
dimensional situation as in a thin plate, or three
dimensional as with a bar. For a pulse traveling in
a thin plate

$$\sin \beta_2 = c_2/c_p = \left[(1 - \nu)/2\right]^{\frac{1}{2}} \qquad (8.12)$$

where c_p is the velocity of long pressure pulses in
a thin plate. For a pulse traveling along a bar

$$\sin \beta_2 = c_2/c_o = \left[2(1 + \nu)\right]^{-\frac{1}{2}} \qquad (8.13)$$

at a considerable distance down the bar. Values of
β_2 for thin plates are plotted against Poisson's
ratio in Fig. 8-7.
    Two quantities can be observed in thin plates

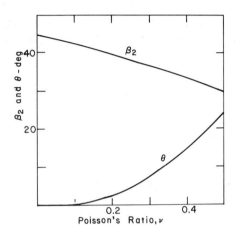

Fig. 8-7.   Upper curve: angle of emergence β_2 of a
            shear wave generated in a plate by glanc-
            ing incidence as function of Poisson's
            ratio. Lower curve: angle θ by which
            stress ellipsoid is rotated by super-
            position of reflected longitudinal pulse
            and incident pulse as function of
            Poisson's ratio.

using photoelastic techniques. These are θ, the
angle by which the stress ellipsoid is rotated by
the superposition of the reflected longitudinal
pulse and the incident pulse, and the ratio τ_1/τ
where τ_1 is the maximum shear stress of the pulse
formed by superposition of the reflected and longitudinal pulses, and τ is the stress in the newly
generated shear pulse. The angle θ, which is given
by

$$\tan 2\theta = 2\nu^2/(1 - \nu)(1 - \nu^2)^{\frac{1}{2}} \qquad (8.14)$$

is plotted in Fig. 8-7 as a function of ν. The ratio
τ_1/τ is plotted as a function of ν in Fig. 8-8 and
is given by

$$\tau_1/\tau = \left\{\nu^2 + [(1 - \nu)/\nu]^2(1 - \nu^2)/4\right\}^{\frac{1}{2}}. \qquad (8.15)$$

    Comparison of the three curves in the two fig-
ures shows clearly that as the angular rotation of

the stress wave ellipsoid becomes nearly equal to
the angle of emergence of the shear wave, the shear
wave becomes relatively much stronger. This is
intuitively reasonable since increasing the rotation
of the ellipsoid implies displacements which are more
nearly parallel to the shear wave front.

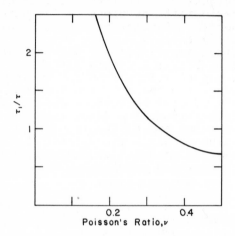

Fig. 8-8.   Ratio of the maximum shear stress of the
             pulse formed by superposition of reflected
             and longitudinal pulses, τ_1, to the stress
             in the newly generated shear pulse, τ, as
             function of Poisson's ratio.

WAVES IN CYLINDRICAL BARS

     The nature of the waves developed in a cylin-
drical bar as a result of applying an impulsive load
to one end has been studied both theoretically and
experimentally. The mathematics is fairly heavy in
the theoretical treatment but generally the solutions
describe the observations reasonably well. Fortun-
ately, it is possible to predict the main features of
the waves using the rather simple geometrical
approach taken for the development of waves in
plates.
     The pattern of wavelets and shear wave segments
established in a cylindrical bar is in many respects
similar to that established in a plate. The princi-

pal difference is that a nonconvergent wave develops at each surface of the plate whereas in a
clyinder, whose single concave boundary focuses the
newly generated expansion and shear wave, high stress
gradients are caused to develop near the axis.
Viewed in cross section, however, the wave patterns
in the plate in Fig. 8-3 and in the cylinder will
appear identical. In the cylinder, as with the
plate, a periodicity develops in the wave form which
depends upon the diameter of the cylinder and its
elastic constants.

    A typical experimentally observed wave form is
reproduced in Fig. 8-9. This was obtained by mount-
ing a strain gage 1.51 m from the end of a 3.8 cm
diameter magnesium bar. A step load was applied by
allowing a mild air shock to strike the end of the
rod and reflect from it. Under these conditions the
stress across the end of the rod increases to a
constant value within a microsecond and remains
constant over an interval of nearly a millisecond.

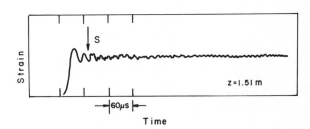

Fig. 8-9.   Strain gage record obtained in 3.8 cm
            diameter magnesium rod at distance 1.51 m
            from end which had been subjected to mild
            air shock. S indicates onset of higher
            frequency oscillations.

    With the arrival of the head of the pulse down
the bar, the strain increases rapidly but not
abruptly to its maximum value. The relatively slow
rise is a consequence of energy having been extracted
from the parent dilatation wave (AB in Fig. 8-3) by
the expansion wavelets and the trailing shear wave.
The rise time depends upon Poisson's ratio and on the
radius and length of the bar. When $\nu = 0.29$ with a
bar 1.3 cm in diameter and 10 cm long, the observed

rise time was of the order of 2 μsec. With a bar
2.54 cm in diameter, 68 cm long, it is of the order
of 3 μsec, and with a bar 2.54 cm in diameter, 180 cm
long, about 5 μsec.

The initial rapid increase in stress is followed
by a few oscillations having the appearance of a
damped sine wave with a gradually decreasing period.
This decrease in period or increase in frequency is
caused by slowing down of the longitudinal components of the wave which increases angle θ, the
obliquity of the shear wave front. Shortly, near
the time indicated by the arrow at S (Fig. 8-9), the
regularity is interrupted by the superposition of a
second set of much higher frequency oscillations.
These herald the arrival of the tail of the pulse,
IJ in Fig. 8-3. Since the tail is moving much slower
than the head of the pulse, the ar_ival of these high
complex frequencies is, as illustrated in Fig. 8-10,
delayed more and more as the pulse moves down the
bar. Beyond S, the oscillations are quite complex

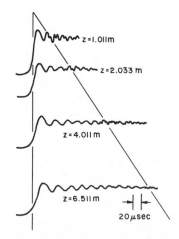

Fig. 8-10.   Series of records taken at increasing
             distance, z, along rod of Fig. 8-9.
             Time scale is horizontal.

but resolve themselves again into a few relatively
simple vibrations with a definite period as the tail
passes. Sustained lateral vibrations of the bar are
felt.

The frequency of the obscillations that arrive immediately after the arrival of the head can be calculated from Eq. (8.8). Some experimental results which are plotted in Fig. 8-11 alongside theoretical curves calculated from Eq. (8.8) show that the agreement between theory and experiment is quite good.

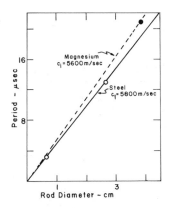

Fig. 8-11. Period of oscillations of first arrivals (Fig. 8-9) as function of rod diameter. Dashed line: theoretical values for magnesium; solid line: for steel. Open circles experimental results for steel and closed circle for magnesium.

Experiments show that after a distance of about twenty times the diameter of the bar, the pulse becomes almost stable in form and the interactions of the several propagating, reflecting, and interacting waves and wavelets has presumably reached a kind of equilibrium state.

MEASUREMENT OF ELASTIC CONSTANTS

Generally a single input pulse applied to the end of a cylinder gives rise to several pulses at the other end. The observed respective arrival times of these pulses, which in accordance with Eq. (8.9) depend upon the dimensions of the bar and

and the dilatation and distortion wave velocities, can be used to determine elastic constants. Usually the input pulse is generated by means of a piezo-electric crystal and a similar crystal is used for the detector at the far end. Some typical records for a series of rods of the same diameter, 2.540 cm, but of different lengths are shown in Fig. 8-12. The A pulse is the first to be received and its time of arrival can be used to calculate c_1. The A pulse is followed by a B pulse that can then be used to fix c_2.

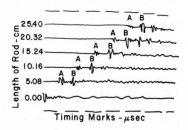

Fig. 8-12.   Strain records obtained at ends of series of cold rolled steel rods all 2.540 cm in diameter which had been subjected to impulsive loading.

CHAPTER 9

SHOCK WAVES IN SOLIDS

INTRODUCTION

Only materials that behave in a purely elastic manner have been considered thus far. Elastic materials, because of their rigidity, transmit both waves of dilatation and waves of distortion. When the intensity of a transient stress is very high, a material may lose its rigidity and react hydrodynamically to the stress or pressure as if it were a fluid. Only one type of wave is possible in a fluid, a compressive longitudinal or dilatation. It can be shown that the velocity c_3 of such a wave is given by

$$c_3 = (dP/d\rho)^{\frac{1}{2}} \qquad (9.1)$$

where P is the pressure level of the wave and ρ is the density of the fluid. In normal fluids at low pressures, c_3 is essentially constant. But at higher pressures, the pressure density curve bends upward causing c_3 to increase with increasing pressure as a result of the increase in slope, $dP/d\rho$. Increments of high pressure will therefore propagate faster than those of low pressure. The result is the evolution of a shock wave, a discontinuity in density, pressure, and temperature which advances through the material with a velocity corresponding to the maximum pressure in the pulse.

It is easy to see how a shock wave develops. Suppose that as a result of a rapid displacement that might be generated by impact or an explosive charge, a plane pressure wave begins advancing through a material. Assume that near the beginning it has the form shown in Fig. 9-1a. Compression starting in the positive direction at point a will appear to travel with velocity c_a relative to the

material at the point. The speed of the point a with
respect to a stationary observer is $c_a + u_a$ where u_a
is the particle velocity at point a. Similarly, a
compression at point b will appear to travel with a
speed $c_b + u_b$ relative to the stationary observer.
When the pressure at b is greater than at a, the
velocity of an element of the wave at b will be
greater than at a and the disturbance will advance
faster at b than at a. The regions of higher pressure
in the wave will approach the slower moving lower
portions ahead of it and the situation shown in
Fig. 9-1b will develop. The ultimate result of this
overtaking effect will be to make the front of the
wave very steep as shown in Fig. 9-1c.

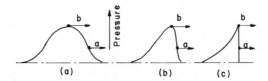

Fig. 9-1. Stages during the development of a shock
wave. Disturbance advances faster at
point (b) than at point (a).

It is obvious that large temperature and pressure gradients will develop at the steep front, with
large amounts of energy being dissipated, effects
that were neglected in the development of the elastic
wave equations. Results based on the elastic wave
theory cannot be applied directly to steep fronts.
The slope of the front obviously cannot be infinitely
steep since this would imply infinite accelerations.
Shock waves have very short rise times, not uncom-
monly a few nanoseconds.

Tensile shock waves cannot develop. Such waves
in which the latter portions are regions of lower
pressure and the particle velocities are away from
the direction of advance, will broaden out rather
than shock up.

In most practical problems involving impact and
explosions, both elastic waves and shock waves enter
the picture. Sometimes an initial shock wave transforms itself into an elastic wave as its energy is
dissipated, or it can become the source of an elastic
precursor. The interplay of the waves depends in a

large measure on the intensity of the pressure, the
reaction of the material to it, and the relative
values of shock wave and elastic wave velocities.
At fairly low pressures, the quantity $(dP/d\rho)^{\frac{1}{2}}$,
applicable to shock waves, is less than
$[3K(1-\nu)/\rho(1-\nu)]^{\frac{1}{2}}$, applicable to elastic waves, so
that an elastic precursor wave will develop. At
high pressures, just the reverse will be true with
the shock wave outdistancing the elastic disturbance.
     Using techniques to be described later in this
chapter, shock pressures as high as 15 megabars can
be attained. Such techniques have made possible the
accumulation of extensive data on how materials behave when subjected to pressures simulating the
extreme pressure conditions existing within the
interior of the earth.
     Shock waves and their effects is a full subject
in its own right and cannot be treated extensively
here. This brief introduction to shock wave theory
is intended to complement the discussions on transient elastic waves, and especially to bring out the
similarities and differences between the two types
of waves. Generally, both types are similar when the
situation is one dimensional, the waves are plane,
and traveling in a medium of infinite extent. It is
when the waves are divergent or interact with boundaries that significant differences appear. These
differences are particularly apparent for oblique
incidence where in the case of elastic waves much of
the energy of the dilatation wave can be transformed
into shear wave energy.

                 CONDITIONS AT A SHOCK FRONT

     Experience and qualitative considerations con-
firm that for most practical purposes the thickness
of a shock front is negligible and that in contrast
to the changes occurring behind it, the front can be
looked upon as a discontinuity in pressure, density,
and temperature. The equations applying to such a
discontinuity were originally developed by Rankine
and by Hugoniot. They are easily developed by considering regions immediately ahead of and behind the
discontinuity.
     Consider Fig. 9-2. If an observer moves with
the velocity U of such a front into a region of
particle velocity u_o, which in many situations is

equal to zero, and density ρ_o, the apparent velocity of the fluid toward him is $(U - u_o)$. In a time dt, a mass of material, $\rho_o(U - u_o)dt$. will enter per unit area of the front. It is imperative to establish a well defined point of reference. Here the observer can be thought of as moving along with the shock front and it is what he, not an outside observer sees that is of concern. The moving observer sees material leaving the front, receding from him with

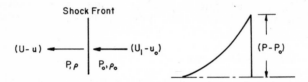

Fig. 9-2.   Physical parameters associated with
            progress of a shock wave.

velocity $-(U - u)$, where the velocity is taken as positive to the right and u is the particle velocity relative to fixed coordinates. The mass of material leaving the front in time dt will be $\rho(U - u)dt$, where ρ is the density of the material to the left of the front. If the front is discontinuous, the time dt can be shrunk to an infinitesimal value for which the mass flow into the front must approach that away from it. In the limit, the two become equal so that

$$\rho_o(U - u_o) = \rho(U - u), \qquad (9.2)$$

an expression of conservation of mass.
        The mass into the front in time dt has momentum

$$\rho_o(U - u_o)u_o dt$$

and the mass flow out has momentum

$$\rho_o(U - u_o)u dt.$$

The difference, which is the change in momentum, must equal the impulse of the new force per unit area in the limit $dt \to 0$. If the pressure ahead of and behind the front are, respectively, P_o and P, then

$$\rho_o(U - u_o)(u - u_o) = P - P_o, \qquad (9.3)$$

an expression of conservation of momentum.

A similar argument with respect to energy rather than momentum requires that the new work done by the pressures P and P_o equal the increase in kinetic plus potential energy when the time increment dt becomes infinitesimal. The work done by P per unit area of front is Pudt, and by P_o is $P_o u_o dt$. The kinetic energies per unit mass are, respectively, $(1/2)u^2$ and $(1/2)u_o^2$. Letting E and E_o denote the respective final and initial internal energies per unit mass, then

$$Pu - P_o u_o = \rho_o (U - u_o)[E - E_o + (1/2)(u^2 - u_o^2] \quad (9.4)$$

expresses the requirements for energy conservation.

In most cases of practical interest, the material ahead of the shock front is at rest, $u_o = 0$. With this simplification and some rearrangement, the above equations can be written in the more convenient form

$$\rho(U - u) = \rho_o U \qquad (9.5)$$

$$P - P_o = \rho_o Uu \qquad (9.6)$$

$$E - E_o = (1/2)(P + P_o)(1/\rho_o - 1/\rho). \qquad (9.7)$$

These equations are equally valid for both plane and spherical shock waves.

If the equation of state, which is volume as a function of stress or pressure, is known for a material, it is possible to determine the increase in internal energy, $E - E_o$, as a function of the pressures P and P_o and the densities ρ and ρ_o. For given values of P_o and ρ_o in the undisturbed material ahead of the shock front, Eq. (9.7) then establishes a relation between P and ρ immediately behind the front. This pressure-density relationship is usually known as the Rankine-Hugoniot equation, or simply the Hugoniot. A graph of it looks somewhat like an adiabatic or isothermal P-V curve, but it is not quite the same function.

Eqs. (9.5) and (9.6) can be solved for the shock velocity U and particle velocity u in terms of the pressures and densities behind the front, giving

$$U = \left[(\rho/\rho_o)(P - P_o)/(\rho - \rho_o) \right]^{\frac{1}{2}} \qquad (9.8)$$

and
$$u = (\rho - \rho_o)U/\rho. \qquad (9.9)$$

Eq. (9.8) is useful when the equation of state, usually obtained experimentally, is known. The equation permits calculation of shock front velocity U in terms of pressure P behind the shock front for the initial conditions P_o and u_o. Similarly, the particle velocity u can be calculated using Eq. (9.9).

Considered generally, Eqs. (9.8) and (9.9) provide three relationships among four variables, P, ρ, U, and u, by means of which the three variables can be expressed in terms of the fourth, given the initial conditions, P_o and ρ_o.

It is apparent from Eqs. (9.8) and (9.9), particularly Eq. (9.8), that a simultaneous experimental determination of shock velocity and particle velocity is sufficient to establish a point on the Hugoniot ρ-v curve and that a series of measurements under varying conditions will define the entire curve.

Extensive single Hugoniot measurements on a large number of substances indicate that for almost all of them, shock velocity and particle velocity are linearly related. The physical reason for this linear relationship:

$$U = a + bu \qquad (9.10)$$

where a and b are constants characteristic of the material, is not understood. It holds, however, for ionic, molecular, and metallic crystals and includes solids and alloys. Sand is a notable exception. A specific linear relation holds only for a single phase. When a material undergoes a phase change, the slope changes at the pressure where the phase change occurs. This fact is used to discover and to locate more precisely where phase transitions occur. Such transitions have been observed in bismuth, granite, iron and steel, marble, playa, pyrolytic graphite, taconite, and tuff, among others.

When shock velocity and particle velocity are linearly related, the equation of state can be written explicitly in terms of the constants a and b of Eq. (9.10). Substituting in Eq. (9.10), the expression for U of Eq. (9.6) yields

$$P = \rho_o u(a + bu) \qquad (9.11)$$

when P_o, usually equal to one atmosphere, is considered negligibly small compared to P. Eq. (9.11) can then be written in the form

$$v/v_o = [a + (b - 1)u]/(a + bu). \qquad (9.12)$$

Eliminating u between Eqs. (9.11) and (9.12) gives

$$P = \rho_o a^2 \eta / (1 - b\eta)^2 \qquad (9.13)$$

where

$$\eta = 1 - (v/v_o). \qquad (9.14)$$

The equation of state, expressed by Eq. (9.13), is extremely useful in computing thermodynamic quantities. It should be noted, however, that it is applicable only when shock velocity and particle velocity are linearly related. Numerous investigators have expressed their experimental results in the form of Eq. (9.13) although the data can also be expressed, and often is, in the purely empirical and analytic form

$$P = A\mu + B\mu^2 + C\mu^3 \qquad (9.15)$$

where

$$\mu = (\rho/\rho_o) - 1 \qquad (9.16)$$

and A, B, and C are material dependent constants.

EXPERIMENTAL METHODS FOR OBTAINING EQUATION OF STATE DATA

There has been a rapid accumulation of data pertaining to the behavior of materials, metals, plastics, liquids, and ionic compounds subjected to intense dynamic loading. Much of these data relate to the dynamic equation of state or Hugoniot. The experimental results have provided the constants needed to fix in a quantitative fashion the thermodynamic parameters associated with compression under shock loading.

Empirical data used in determining Hugoniot curves have been obtained principally by making velocity measurements. In most of the early work, shock velocity and particle velocity were measured simultaneously and the conservation equations were used to compute pressure-volume relationships and other thermodynamic constants. Later, after Hugoniots had been well established for some materials, the impedance match method became more popular since it involved only the determination of shock velocity. Some attempts have been made to measure density changes directly using flash X ray techniques to determine density, but in general, these have not given very accurate results. Direct measurements of pressure are being made successfully at present using piezoelectric quartz crystals, although the upper limit pressure by this technique is only about 40 kilobars.

Several methods have been used to generate shocks. In early tests, an explosive charge was detonated in intimate contact with the material under study using a special explosive charge that generated a plane shock wave. A severe limitation of this technique is that a wide variation of shock pressures cannot be achieved. An important modification of the method was the use of a driver plate propelled by an explosive charge so as to strike the specimen. By judicious choice of driver plate material and explosive charge size, a very wide range of pressures was possible.

More recently, a number of laboratories have developed gun impactor-plate devices for generating shocks. The big advantage with these is the possibility of accurately preselecting and controlling initial conditions. With guns, extremely high pressures, several thousand kilobars, are possible.

Pin contactors to measure free surface velocity gave the first quantitative data on particle velocity, the assumption being made that free surface velocity was just twice that of particle velocity. The technology of the pin contactor has reached an exceedingly high level of development although other techniques are gradually replacing this one. One such technique utilizes the fact that argon becomes luminescent when subjected to high intensity shock, making it possible to measure times of arrival by observing onsets of luminosity with a streak camera. In another technique, surface velocity is monitored continuously by means of a resistance wire. Condenser techniques have also been found useful.

THEORY OF HYDRODYNAMIC IMPACT

Consider the hypothetical case of two semi-
infinite bodies colliding along a plane interface,
one body, medium 1, moving with velocity, V, in a
direction perpendicular to the interface (Fig. 9-3);
the other, medium 2, is stationary. Plane shocks
will be propagated from the interface into both
colliding bodies as indicated in the right hand drawing. For most practical as well as theoretical
purposes, each shock front may be considered a zone
of infinitesimal width across which there is a discontinuous jump of pressure and velocity of the
medium.

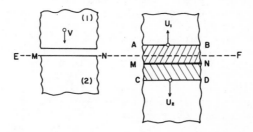

Fig. 9-3.   Shock waves developed during impact of two
            plates.   Left: before impact; right:
            after impact.   Impact velocity, V.

The following relationships have been derived
for the changes across the shock front, propagated
into a body at rest, from the laws of conservation of
mass, momentum, and energy:

$$U\rho_o = (U - u)\rho \qquad (9.17)$$

$$P = \rho_o U u \qquad (9.18)$$

and

$$E = P/2(1/\rho_o - 1/\rho) \qquad (9.19)$$

where U is shock velocity; u is particle velocity
behind the shock front; ρ_o is the initial density;
ρ is the density behind the shock front; P is the
change in pressure across the shock front; and E is

the change in internal energy across the shock front. These conditions must hold at all times during the course of the impact.

Two boundary conditions further connect the shocks in the two bodies: because the two bodies must remain in contact during the collision, the velocities of the two materials on both sides of the interface must be the same, which is the boundary condition of continuity of particle velocity; and secondly, from Newton's third law, the pressures in the two shocks must be equal:

$$P_1 = P_2. \qquad (9.20)$$

Viewed from coordinates fixed with reference to the interface, MN, the particle velocity between the two shocks is zero: the material on each side of the interface appears to an observer, moving with the interface, to be at rest, with the shock fronts, AB and CD, moving out into each respective medium at a velocity determined by momentum considerations. In homogeneous media, the shock velocities will remain constant.

Consider now what happens to the several planes: AB, the front of the shock moving upward into medium 1; MN, the plane of common contact between medium 1 and 2; and CD, the front of the shock moving downward into medium 2. EF is a fixed plane of reference, at impact being coincident with MN. After unit time, MN will have moved down from EF a distance u, u being the particle velocity in the shock waves; CD will have moved a distance U_2 into medium 2 from EF and will be a distance $(U_2 - u)$ from MN, U_2 being the shock velocity in medium 2; and AB will have moved a distance U_1 into medium 1 and will lie at a distance $(U_1 - v)$ upward from EF, where U_1 is the shock velocity in medium 1. AB will lie a distance $(U_1 - V)+u$ from MN.

Look now at the compression of the two bodies:

$$\delta_1 = (\rho_1 - \rho_{10})/\rho_1 = (v_{10} - v_1)/v_{10} \qquad (9.21)$$

$$\delta_2 = (\rho_2 - \rho_{20})/\rho_2 = (v_{20} - v_2)/v_{20} \qquad (9.22)$$

where δ_1 is the compression of medium 1; δ_2 is the compression of medium 2; ρ_{10} and ρ_{20} are the original densities of mediums 1 and 2, respectively; ρ_1 and ρ_2 are densities of compressed mediums 1 and 2, respectively; and the v's are specific volumes.

The mass, m_1, of medium 1 which before impact was contained in the volume U_1, after unit time resides in volume $(U_1 - V + u)$; and the mass, m_2, of medium 2, originally residing in the volume U_2, now resides in volume $(U_2 - u)$. Thus, since by definition

$$\rho_1 = m_1/(U_1 - V + u) \qquad \rho_{10} = m_1/U_1 \qquad (9.23)$$

$$\rho_2 = m_2/(U_2 - u) \qquad \rho_{20} = m_2/U_2 \qquad (9.24)$$

Eqs. (9.21) and (9.22) lead to

$$\delta_1 = \left[m_1/(U_1 - v + u) - m_1/U_1 \right]/\left[m_1/(U_1 - V + u) \right] \qquad (9.25)$$

$$\delta_2 = \left[m_2/(U_2 - u) - m_2/U_2 \right]/\left[m_2/(U_2 - u) \right] \qquad (9.26)$$

which reduce to

$$U_1 = (V - u)/\delta_1 \qquad (9.27)$$

and

$$U_2 = u/\delta_2. \qquad (9.28)$$

From conservation of momentum

$$m_1 V = u(m_1 + m_2) \qquad (9.29)$$

so that

$$m_2 = m_1(V - u)/u. \qquad (9.30)$$

By definition and substitution from Eqs (9.27) and (9.28) it follows that

$$m_1 = U_1 \rho_{10} = (V - u)\rho_{10}/\delta_1 \qquad (9.31)$$

and

$$m_2 = U_2 \rho_{20} = u\rho_{20}/\delta_2. \qquad (9.32)$$

Combining Eqs. (9.29), (9.30), and (9.31) and solving for V yields

$$V = u[1 \pm (\rho_{20}\delta_1/\rho_{10}/\delta_2)^{\frac{1}{2}}] \qquad (9.33)$$

and as u cannot exceed V under compression, only the positive root has physical significance. Solving for u gives

$$u = V/[1 + (\rho_{20}\delta_1/\rho_{10}\delta_2)^{\frac{1}{2}}] \qquad (9.34)$$

and using Eq. (9.28) gives

$$U_2 = V/\delta_2[1 + (\rho_{20}\delta_1/\rho_{10}\delta_2)^{\frac{1}{2}}]. \qquad (9.35)$$

Now from Eq. (9.18)

$$P_2 = \rho_{20}U_2u_2 \qquad (9.36)$$

where u_2 and U_2 are, respectively, particle and shock velocities measured with respect to the unshocked material. In the original frame of reference, medium 2 is initially at rest so that $u_2 = u$, the velocity with which the interface between the two mediums moves, and Eq. (.36) becomes

$$P = \rho_{20}U_2u \qquad (9.37)$$

since $P_2 = P_{20} = P$. Note, however, that it is not true that particle velocity, v_1, in medium 1, measured with respect to the unshocked medium, is equal to u. Rather,

$$u_1 = V - u \qquad (9.38)$$

and

$$P = \rho_{10}U_1u_1 = \rho_{10}U_1(V - u). \qquad (9.39)$$

Use of Eq. (9.37) leads finally to

$$P = (\rho_{20}/\delta_2)\{v/[1 + (\rho_{20}\delta_1/\rho_{10}\delta_2)^{\frac{1}{2}}]\}^2 \qquad (9.40)$$

which becomes

$$P = \{v/[(\delta_2/\rho_{20})^{\frac{1}{2}} + (\delta_1/\rho_{10})^{\frac{1}{2}}]\}^2. \qquad (9.41)$$

Eqs. (9.35) and (9.41) permit calculations of shock velocity U_2 and contact pressure P for a given impact velocity V, provided the respective equations of state of the two mediums are known.

On the other hand, by measuring V, the velocity of impact, u, the particle velocity at the interface, and U_2, the velocity of the shock in the impacted medium, Eqs. (9.35) and (9.41) contain only two unknowns, δ_1 and δ_2, and hence can be used to compute an equation of state.

If the impact is between two like materials, then from Eq. (9.34) u = V/2, that is, particle velocity or interface velocity is exactly one half the velocity of impact, and Eq. (9.41) becomes

$$P = (\rho_{20}/\delta_2)(V/2)^2 \tag{9.42}$$

or

$$P = (1/v_{20})[1 - (v/v_{20})](V/2)^2. \tag{9.43}$$

Now employing the condition

$$P = \rho_{20}uU_2 = \rho_{10}u_1U_1 = \rho_{20}(V - u)U_1 \tag{9.44}$$

where u_1 is the particle velocity in medium 1 behind the shock and U_1 is the shock velocity in medium 1, both velocities relative to the unshocked medium 1, that is, $u_1 = V - u$, it can be shown by substitution that

$$V = u_1[1 + (\rho_{10}\delta_2/\rho_{20}\delta_1)^{\frac{1}{2}}] \tag{9.45}$$

and

$$V = \delta_1U_1[1 + (\rho_{10}\delta_2/\rho_{20}\delta_1)^{\frac{1}{2}}]. \tag{9.46}$$

DRIVER PLATE TECHNIQUE FOR GENERATING HIGH PRESSURE

The projection of a plate, either from a gun or by high explosive devices, against a target specimen material is a common method for obtaining intense high pressures. On impact with the specimen , the driver plate or impactor, gives up its momentum in less time than it took to receive it so that the shocks produced in the specimen and the driver plate, which are of equal pressure, are of higher amplitude than the amplitude of the shock produced in the driver plate during its acceleration. Pressures well in excess of 1,000 kilobars can be produced in this way.

For research purposes, the driver plate technique
has the important advantage that different levels of
shock pressure can be produced readily by projecting
the plate at different velocities.

If the Hugoniot curves for the driver plate
material and the specimen material are known, and if
the driver plate velocity is known, and this can
usually be measured during the experiment, then the
shock pressure in the specimen can be readily calculated. The collision of the driver plate and the
specimen produces two shock waves that travel in
opposite directions. The momentum equations can be
written

$$P_1 = \rho_1 U_1 u_1 \qquad (9.47)$$

and

$$P_2 = \rho_2 U_2 u_2 \qquad (9.48)$$

where the subscripts 1 and 2 refer, respectively, to
the driver plate and the specimen. Immediately after
the collision, the conditions at the driver-specimen
interface require that

$$P_1 = P_2 = P \qquad (9.49)$$

and

$$u_1 + u_2 = V_P, \qquad (9.50)$$

where V_P is the velocity of the driver plate.

The shock pressure, P, can now be readily determined graphically from a plot of the type shown in
Fig. 9-4. Here previously determined Hugoniot
equation of state data are used to plot particle
velocity versus pressure for both the driver plate
(upper curve) and the specimen (lower curve). A
pressure, P, is then established, by approximation,
such that $u_1 + u_2 = V_P$. This pressure will be the
shock pressure in both the plate and the specimen.

The solution is easier to obtain when the
driver plate and specimen are of the same material.
There will be only one particle velocity versus
pressure curve and the pressure during impact will be
that corresponding to a point along the curve where

$$u = V_P/2. \qquad (9.51)$$

Collision with the driver plate produces in the
specimen a flattopped pressure pulse which has a
constant amplitude to a depth of at least several
times the thickness of the driver plate.

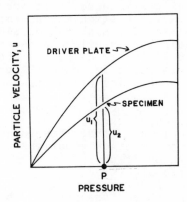

Fig. 9-4.   Particle velocity versus pressure curves
            used to determine pressure, P, generated
            during impact.

IMPEDANCE MATCH METHOD FOR DETERMINING HUGONIOT

When a shock wave encounters an interface between
two dissimilar materials, as indicated in Fig. 9-5,
two new waves will be generated, a transmitted shock
wave and a reflected wave. The fact that the relative

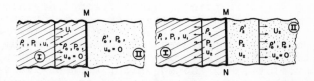

Fig. 9-5.   Physical parameters associated with shocks
            developed in impedance match method of
            determining Hugoniots. MN is interface
            between specimen 1, known Hugoniot, and
            specimen 2, unknown Hugoniot. Left, before
            shock reaches interface; right, shortly
            after.

intensities of these new waves are governed by the
respective compressibilities and densities of the two
interacting materials has been used extensively by
experimental investigators to establish Hugoniot
curves. The method is known as an impedance match
method. The basic stratagem is to generate a shock
of known or measurable strength in a material whose
Hugoniot curve is well established, allow the shock
to be reflected at an interface between the known
material (medium 1 in Fig. 9-5) and the material for
which the Hugoniot is being sought (medium 2), and
then measure the velocity of the transmitted shock.
This procedure is repeated for shocks of several
strengths in order to obtain the points needed to
trace a full Hugoniot curve.

The basis of the method lies in judicious application of the conservation equations and appreciation
of the boundary conditions. At the interface, two
boundary conditions must be met: continuity of
pressure, and continuity of particle velocity. The
system of reflected and transmitted shocks which
develops after the shock reaches the interface is
illustrated in the right drawing of Fig. 9-5. The
pressure, P_2, at the interface is the pressure of the
transmitted shock, and at the same time represents
the sum of the pressure, P_1, of the incident wave and
P_1', the pressure of the reflected wave. The pressure
P_1' may be either positive or negative, depending upon
the impedance match between the two materials. The
situation at the interface can be defined by the
point (P_2, u_2) on a pressure versus particle velocity
diagram. Each shock of a different strength locates
a new point and the locus of all such points defines
the unknown Hugoniot. The problem is to locate each
of the (P_2, u_2) points. Three pieces of information
are sufficient to establish any one point such as
point M in Fig. 9-6: the pressure, P_1, of the
incident shock, the Hugoniot curve for medium 1, and
the velocity of the shock transmitted into medium 2.
The pressure, P_1, the Hugoniot, and conservation
equations fix the point S which has the coordinates
of P_1 and u_1. A curve, the reflection Hugoniot or
cross curve, is a mirror image about the point S of
the P-u curve or known Hugoniot for medium 1 and
portrays possible states of medium 1 with respect to
the state (P_1, u_1). The point M, representing the
state (P_2, u_2), must lie on the curve and must also
lie on the line

$$P = \rho_o U_2 u \qquad (9.52)$$

where U_2 is the velocity of the shock transmitted in medium 2.

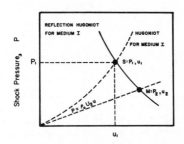

Fig. 9-6.   Graphical method used to determine Hugoniot for unknown specimen in impedance match method of Fig. 9-5.

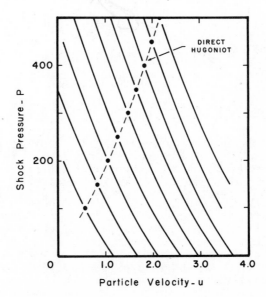

Fig. 9-7.   Generally adopted shock pressure versus particle velocity curves for 24S-T aluminum.

This relationship follows from application of the conservation equations. With ρ_o and U_2 both known, the line can be drawn and its intersection with the reflection Hugoniot locates M.

One of the best established Hugoniots, and most frequently used as a standard, is that for 24S-T aluminum. A number of cross curves for this material are given in Fig. 9-7.

CALCULATION OF TEMPERATURES ASSOCIATED WITH PASSAGE OF SHOCK WAVES

Unlike an elastic wave in which the strain is reversible and adiabatic, a shock wave will leave heated material in its wake. The temperature, T_H, behind the shock front during passage of the wave is calculated from the equation

$$T_H = T_o \, e^{\gamma(1-v_1/v_o)} +$$

$$+ \, e^{\gamma(v_1/v_o)} \int_{v_o}^{v_1} \left\{ (1/2)\left[(dP/dv)(v_o-v) + P\right] \right.$$

$$\left. \cdot \, e^{\gamma(v/v_o)}/C_v \right\}_{HUG} dv \qquad (9.53)$$

where γ is Grüneisen's constant given by

$$\gamma = (dP/dT)_v (v_o/C_v) \qquad (9.54)$$

where C_v is the specific heat at constant volume. The integration is performed numerically along the Hugoniot curve.

The equation is exact but the variation of C_v and $(\partial P/\partial T)_v$ with volume is not known. In most calculations these are assumed constant.

The calculation of the residual or final temperature, T_A, after passage of the shock is made utilizing the relationship

$$T_A = T_H e^{(\partial P/\partial T)_v (1/C_v)(v_H-v)} \qquad (9.55)$$

$$= T_H e^{\gamma\left[(v_H/v_o)-(v/v_o)\right]} \qquad (9.56)$$

where T_H and v_H are the known conditions at any point, taken here as the Hugoniot point. To fix the final temperature and specific volume of the specimen material, the known relation

$$(v - v_o)_{P=0} = v_o \alpha (T - T_o)_{P=0} \qquad (9.57)$$

along the P = 0 isobar is used. Here T_o and v_o refer to the temperature and specific volume and α is an average value of the thermal coefficient of volume expansion.

EMPIRICAL EQUATIONS

Hugoniot data are usually summarized in the form of empirical equations which are especially useful in making thermodynamic computations. Some of the analytical forms are particularly adaptable to computer calculations. Others have a more theoretical basis, their derivation depending upon the empirical linear relationship between shock velocity and particle velocity and are sometimes less convenient for making computations. In any particular case, it is easy to locate an appropriate equation in the extensive literature published in the field of shock wave studies.

Chapter 10

Transfer of Momentum

INTRODUCTION

An impulsively applied blow introduces momentum
into the system to which it is applied. Momentum is
similar to energy in that it cannot be destroyed
but it has the added advantageous quality that it
cannot change its identity and can be kept track of
easily. It always appears as mechanical motion
which moves about through a system distributing itself in various ways. In many practical situations,
it is desired to deliver the momentum to some particular component of the system in order to impart
motion to that component, or to do work such as
crushing, fracturing, or deforming. To control and
hence to utilize the distribution and partitioning
of momentum, it is necessary to understand the
mechanics of transfer processes, including the
factors that influence them.
If the duration of the loading is short, the
length of the transient wave will be commensurate
with the dimensions of the components of the system
under study. When this is the case, the effective
transfer processes are quite different than when
the pulses produced by the loads are long compared
to the dimensions of the parts of the system.
Experimentally it is easy to observe relative
motion of parts. Such observations provide the
means for making quantitative measurements of the
stress, particle velocities, and durations involved
in highly transitory events, events lasting in many
instances only a few microseconds.

RELATION BETWEEN PLATE VELOCITY AND STRESS TIME CURVES

It was shown in Chapter 4 that the total

momentum, M, per unit area in a transient stress pulse propagating in a material of density ρ is given by

$$M = \int \rho v(x) dx \qquad (10.1)$$

where the integration is taken over the length of the pulse. In incremental form this can be written in the three equivalent forms:

$$dM = \rho v dx = (\sigma/c) dx = \sigma dt. \qquad (10.2)$$

It follows from this equation that

$$\sigma = dM/dt. \qquad (10.3)$$

Consider a plate of finite thickness, L, and infinite lateral extent placed loosely against another plate, also laterally infinite as drawn in Fig. 10-1. A sharp fronted compressive wave moving

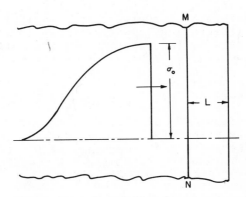

Fig. 10-1. Laterally infinite slab of thickness L affixed loosely along boundary MN to plate of same material. A stress transient is approaching interface from left.

through the left hand plate with its wave front parallel to the interface will enter the right hand plate without change in intensity or form, provided the two plates are of the same material. The wave will then be reflected from the far right face of the right hand plate as a tension wave. It should be noted that whereas the stress changes sign, the

particle velocity does not. In both the incident
and reflected waves, particle velocity is directed
toward the right. When the reflected tension wave
reaches the interface, MN, the interface will separate since it cannot support tension and the right
hand plate will fly off to the right.

The relative amount of momentum trapped by the
right hand plate depends upon its dimensions with
respect to the wavelength of the pulse. If the
thickness of the plate is at least one half the
length of the wave, all of the momentum of the pulse
will be trapped in the plate and its velocity, V,
will be given by

$$V = M/(\rho L/2) = \int v(x)\,dx/(L/2) \qquad (10.4)$$

$$= \int \sigma dt/(\rho L/2). \qquad (10.5)$$

If plate thickness is less than one half the length
of the wave, then the shape of the wave and its
intensity will determine how fast the plate will
move away. For a monotonically decreasing particle
velocity or stress function, thinner plates will
move off relatively faster than thicker plates.
Specifically, the throw-off velocity for the plate
in Fig. 10-1 will be

$$V = (2/\rho L)\int_0^{2L/c_1} \sigma dt = (2c_1/L)\int_0^{2L/c_1} v dt. \qquad (10.6)$$

In a later section it is explained how measurements of the throw-off velocities of a series of
plates can be used to determine the functional
dependence of stress on time, $\sigma(t)$.

IMPEDANCE EFFECTS

At a boundary between dissimilar materials,
part of the wave will be transmitted and part will
be reflected because of their different impedances
to passage of the wave. The proportions of the
reflected and transmitted wave are governed by Eqs.
(7.5) and (7.6). Since the impulse applied to the
right hand plate of Fig. 10-1 is equal to the time
integral of the stress, it follows that the momentum
per unit area, M_T, transferred to a dissimilar

plate is given by

$$M_T = 2\rho'c_1'/(\rho'c_1' + \rho c_1)\int_0^{2L/c_1'} \sigma_I dt. \quad (10.7)$$

Thus for a receptor plate with very high specific
acoustic resistance as compared with the specific
acoustic resistance of the donor plate, there is
nearly a two to one amplification in momentum transfer. Momentum is, of course, conserved, the additional momentum imparted to the receptor plate
appearing as momentum reflected to the left in the
left hand donor plate (Fig. 10-1). This means that
if the left hand plate were of finite thickness it
would, as the result of an impulsively applied
blow, bounce off to the left. If the specific
acoustic resistance of the receptor plate, $\rho'c_1'$,
is less than the specific acoustic resistance of the
donor plate, ρc_1, the donor plate continues to move
to the right.

The velocity, V, with which the right hand plate
will move off is given by

$$V = M_T/\rho'L = (c_1'/L)/(\rho'c_1'+\rho c_1)\int_0^{2L/c_1'} \sigma_1 dt. \quad (10.8)$$

GAPS

Consider the situation shown in Fig. 10-2 in
which there is a narrow free space or gap between a
main body and a receptor plate. Obviously no
momentum can enter the receptor plate until the gap
is closed. Closure will take time, T_2 given by

$$D = 2\int_0^{T_2} v(t)dt. \quad (10.9)$$

where D is the gap width and $v(t)$ is the particle
velocity distribution in the wave approaching the
gap. Until the gap closes, the transient wave is
being reflected from the free surface of the body
back into itself. Eventually the gap may close
because of the rightward motion imparted to the free
surface of the body as the incident wave impinges

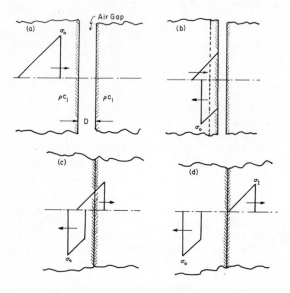

Fig. 10-2.   Mechanics of closure and filtering
            action of an air gap acted upon by a
            stress transient. (a) approach of the
            transient toward gap; (b) reflection
            and beginning of closure; (c) shortly
            after closure; (d) distribution of stress
            at instant interaction with joint is
            complete.

against it. As soon as the gap closes, the situation
changes, momentum enters the receptor plate and
continues to do so until the wave completes its
passage across the interface. This chain of events
is illustrated in the figure for the case of a sawtooth wave in which the gap closes in a time
corresponding to about one half the duration of the
wave.

      In general, the effect of a gap is to attenuate
severely the amount of momentum transmitted to a
receptor plate. No momentum will, of course, be
transmitted if the time to close the gap is equal
to or greater than the duration of the pulse.

MULTIPLE PLATES

When several receptor plates are stacked together as shown in Fig. 10-3, each one will take up part of the momentum and the several plates will assume different velocities depending upon the shape of the transient disturbances and the respective thicknesses of the plates. The velocity V_1 of the plate farthest to the right in the figure will be given by

$$V_1 = (1/2\rho L_1) \int_0^{2L_1/c_1} \sigma(t) dt \qquad (10.10)$$

where L is the thickness of the plate and the stress is given by $\sigma(t)$. The velocity of each succeeding plate will be less than the one before if $\sigma(t)$ is a monotonically decreasing function of time.

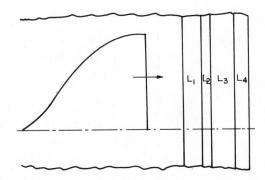

Fig. 10-3. Action of stress transient on juxtaposed loose receptor plates.

Specifically, the velocity V_i of the ith plate will be given by

$$V_i = (1/2\rho L_i) \int_{2L_i - 1/c_1}^{2L_i/c_1} \sigma(t) dt. \qquad (10.11)$$

The situation becomes more complicated when the plates consist of dissimilar materials, for then part of the stress will transmitted and part reflected at each boundary as described earlier. The dynamics of the momentum transfer become complicated fairly rapidly. Usually the best procedure in any particular case is to trace graphically the progress of the wave step by step as it moves through the assemblage of the plates.

A specific example of a graphical solution is illustrated in Fig. 10-4. Here two plates of identical material are separated by a bonding plate which

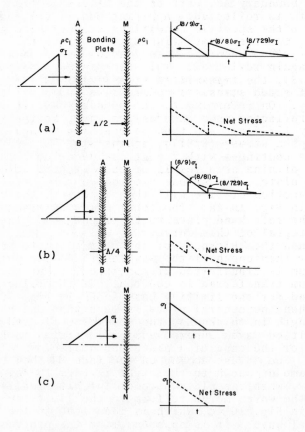

Fig. 10-4. Effect of thickness of bonding plate on reflection and transmission of a stress transient.

has a density just half that of the other two but
which is capable of transmitting a longitudinal wave
with the same velocity. The thickness of the central
plate is assumed to be equal to $\Lambda/2$ where Λ is the
wavelength of the disturbance. The momentum, M,
transferred to the right hand plate will depend upon
the stress, σ_{MN}, exerted against the face MN and will
be given by

$$M = \int \sigma_{MN}(t)dt. \qquad (10.12)$$

At the boundary AB, part of the stress is transmitted
and part is reflected. Applying Eqs. (7.5) and
(7.6) to the specific conditions assumed here gives
$(2/3)\sigma_I$ for the stress transmitted across the
boundary AB. When the wave reaches the boundary MN,
it is again reflected. This time in accordance with
Eq. (7.5), the transmitted stress will be $(8/9)\sigma_I$.
The reflected stress is compressive and equal to
$(2/9)\sigma_I$. On returning to the boundary AB, it is
again reflected as a compressive wave, having intensity $(2/27)\sigma_I$ and then on reaching MN again
develops a stress $(8/81)\sigma_I$ against the surface. The
process continues with the stress decaying by a
factor of nine as a result of each transit across the
middle plate. The stress exerted on the surface MN
as a function of time will then appear as the curve
shown in Fig. 10-4a. The total momentum transmitted
from the left hand plate to the right hand plate is
the integral of this curve.
     When the thickness of the central plate is
greater than one half the wavelength of the pulse,
momentum is transferred discontinuously but the total
momentum transferred is constant, at the value
obtained for the limiting case shown in Fig. 10-4a.
     When the central plate is less than one half
wavelength in thickness, the several reflected and
transmitted waves superpose as illustrated in Fig.
10-4b for the case of a plate one quarter wavelength
thick. The stress acting on the face MN then varies
with time as shown to the right in Fig. 10-4b.
Finally, when the plate becomes infinitesimally
thin, the waves merge to form in the limit the wave
shown in Fig. 10-4c, which is identical to the
original wave. It is apparent from the progression
of the curves with decreasing thickness of the
central bonding plate, that a thin cemented bond
between two identical blocks will not affect appre-
ciably the transmission of momentum.

OBLIQUELY INCIDENT WAVES

When a plane wave strikes a loose boundary obliquely, part of the wave is transmitted and part is reflected. The precise way in which momentum is partitioned depends upon the Poisson's ratio of the material and upon the angle of obliquity of the wave. Since the interface cannot support shear, only normal stresses or particle velocities are transmitted and the plate in Fig. 10-5 must move off in a direction normal to the interface which serves as a filter and allows transmission only of that portion of the momentum which is normal to it. The rest of the momentum is trapped in the first body, causing relative movement along the interface.

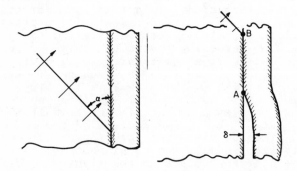

Fig. 10-5.   Dynamics of detachment of loosely affixed plate acted upon by an obliquely incident stress transient.

The relative amount of momentum transferred to the receptor plate can be computed from the intensities of the shear wave F and the dilatation wave E (see Fig. 7-18) that are generated in the plate. Remembering that stress is equal to $\rho c v$, the momentum M, transmitted will be given by

$$M = \int \rho c_2 v_{F_y} dt + \int \rho c_1 v_{E_y} dt \qquad (10.13)$$

where ρ is the density of the material and v_{F_y} and v_{E_y} are, respectively, the particle velocities in the transmitted shear and dilatation waves. The particle velocities for any particular combination of Poisson's ratio and angle of obliquity can be read from curves (Figs. 7-15 and 7-16).

It is apparent that at angles of obliquity ranging
from 40° to 75°, a fairly large fraction of the
momentum will be trapped in the first body and will
never reach the receptor plate.
    Separation of plate from the main body will
occur progressively as the wave moves along the
boundary but not all at once as is the case for
normal incidence. The point A (Fig. 10-5) at which
separation is occurring lags behind the wave front
by the time it takes the front to go obliquely across
the plate and return. Thus distance $\overline{AB}$ is given by

$$\overline{AB} = 2L \tan \alpha \qquad (10.14)$$

where L is the thickness of the plate and α is the
angle of incidence of the wave. The amount of separation and the distortion of the plate are greatly
exaggerated in Fig. 10-5. Generally, particle
velocities are very much lower than wave velocities
so that the separation, δ, is quite small compared to
$\overline{AB}$. If, for example, the plate were a small diameter
pellet resting on a much larger plate, the pellet
would have hardly moved at all by the time the wave
had swept across its face.
    The situation is very different when the bound-
ary between the two plates is cohesive and able to
support shear stresses. The wave will pass through
the boundary unaltered, be reflected from the right
hand face of the receptor plate and return as a
tension wave to the boundary. What happens then
depends upon the strength of the bond between the
two pieces. If it is strong in tension, the wave
passes back on through and the plate remains in
place; if the bond is weak in tension, it will break
and the plate will fly off. In general, if the plate
flies off, it will move not only normal to the interface as was the case with the loose boundary but
will also have a component of velocity along the
interface. When the plate is very thin, its direction of throw will be the same as the direction the
surface of the main body would move if the plate
were not there. This direction is in accordance
with the equation

$$\sin^2 (\overline{\alpha}/2) = [(1/2)(1-2\nu)/(1-\nu)] \sin^2\alpha \qquad (10.15)$$

where $\bar{\alpha}$ is the angle between the trajectory of the plate and the normal to it.

    Figure 10-6 is a diagram illustrating the direction in which small rigidly attached pellets would fly off when a sharp fronted elastic disturbance strikes the free boundary obliquely. It is assumed

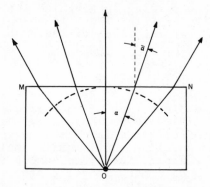

Fig. 10-6.   Diagram illustrating direction in which a small rigidly affixed pellet would fly off when a stress transient strikes a free surface, MN, obliquely.

in this illustration that a spherical wave originates at the point O and expands outward, finally striking the surface MN on which are affixed a few thin pellets. The particle velocity, except for the point on the axis, which is a special case, is not, in general, perpendicular either to the surface or the wave front. Its direction, which near the axis is almost normal to the wave front, deviates more and more with increasing angle of incidence and can be computed from Eq. (10.15). The arrangement described here has been used successfully to determine the dynamic Poisson's ratios of several materials. The throw angle, $\bar{\alpha}$, can be measured and as the angle of incidence, α, of the wave is known, computation of ν from Eq. (10.15) is possible.

EDGE EFFECTS

In most real situations, edge effects must be
taken into account, as the elements involved in the
momentum partitioning and distribution processes are
not plates laterally infinite in extent.

A simple situation is illustrated in Fig. 10-7.
Here a small circular pellet of diameter, d, and
thickness, L, is placed on the surface of a plate
which, for all practical purposes, can be considered
to be laterally infinite. As soon as the wave front
AB reaches the free surface of the plate, the front
will be divided into two parts, one of which is
reflected from that portion of the plate not covered
by the pellet, and the other which enters the pellet.

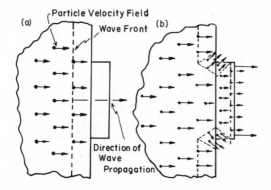

Fig. 10-7. Sketch illustrating certain qualitative
            features of particle velocity fields
            which contribute to edge effects.

An exact quantitative treatment of the situation is
not feasible, but in a qualitative way it can be seen
that soon the pattern of wave fronts and particle
velocity fields develop as illustrated in Fig. 10-7b.
As the longitudinal wave moves through the pellet,
the edge of the pellet begins to expand laterally as
described in Chapter 8. The material in the plate
which lies immediately beneath the pellet will remain
in a state of compression while at the same time that
surrounding the pellet will be relieved by the wave
that is reflected from the free surface of the large
plate. At this time, the material beneath the

pellet will begin to expand into the region that has
been relieved by the reflected wave. In spite of
this expansion, the joint between plate and pellet
will remain intact until the wave front passes
through the pellet and back to the interface. During
this transit of the wave, momentum continues to be
imparted to the pellet. The principal effect of
expansion is to decrease the velocity of propagation
of the rearward portions of the wave, resulting in
somewhat less momentum being transferred to the
pellet than if its diameter were infinite. The
decrease in velocity will be at most twenty per cent,
the change from the higher dilatation wave velocity
to the lower rod velocity.

     To minimize or ignore edge effect uncertainties
when using pellets to measure particle velocities
and stresses, the pellets should be relatively thin.
Certainly the thickness to diameter ratio should
never exceed one. As the thickness increases and
the time increases, the edge effects become relatively more pronounced. Finally, when pellet thickness
reaches pellet diameter, edge effects become
significant.

     Edge effects can have a pronounced influence on
momentum transfer and partitioning. One such
example is shown in Fig. 10-8 where a transient
stress wave enters a boss upon which is affixed a
small pellet. The wave pattern shown in Fig. 10-9

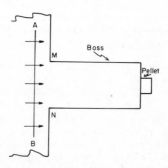

Fig. 10-8. Geometry for analyzing transfer of
            momentum from a stress transient to a
            pellet loosely affixed to a boss.

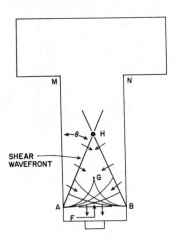

Fig. 10-9.   Wave patterns developed during transfer
            process of Fig. 10-8.

develops.   The pattern contains three distinct
regions.   The region outlined by the surface ABFA,
which flattens rapidly with progress of the wave, is
a region unaffected by the boundary of the boss.
Particle velocity in this region will be directed
along the axis of the boss and be nearly uniform
except near the boundary where the wave is greatly
weakened through supplying energy to the expansion
and shear waves.   The region bounded by the surface
AFBHA is undergoing lateral expansion and in general
contains relatively less forward momentum than the
unaffected region bounded by ABFA.   The shear wave
front AHB is the start of the third region.   The
shear wave will be strong and much of its momentum
is directed forward.
        It is instructive to consider momentum transfer
for three possibilities: one, in which the thickness
of the pellet and the length of the boss are considered fixed and the diameter of the pellet is
varied; two, where the diameter and thickness of the
pellet are held fixed and the length of the boss is
varied; and finally where the length of the pellet
is varied, holding the length of the boss and the
diameter of the pellet fixed.

In considering the first proposition, assume
that the thickness of the pellet, taken as constant,
is less than one half the perpendicular distance
from the line AB (Fig. 10-9). to the point F. Small
diameter pellets will not feel the effect of the
boundary at all. Edge effects will begin to be felt
as soon as the diameter of the pellet becomes large
enough so that some of the surface AFBA enters the
pellet before it has been flung from the boss. In
any particular case, it is a simple geometrical
problem to calculate when the pellet first becomes
affected. Figure 10-10 is schematic plot of velocity
of pellet versus diameter of pellet. Note how the

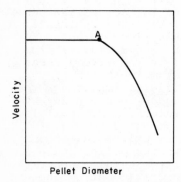
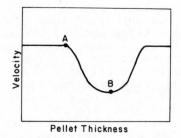

Fig. 10-10. Throw-off velocity of pellet of Fig. 10-8
             as function of pellet diameter. Point A
             is where expansion wavelets begin to
             enter pellet.
Fig. 10-11. Throw-off velocity of pellet of Fig. 10-8
             as function of pellet thickness. Point A
             is where expansion wavelets begin to
             enter pellet and point B is where shear
             wave begins to enter pellet.

velocity of the pellet begins to decrease when its
diameter reaches the critical value, point A on the
curve, at which the surface AFBA enters it. The
break point occurs when the distance between the
front AB and the point F (Fig. 10-9) corresponds to
just twice the thickness of the pellet. As the
pellet increases in diameter, in addition to being
affected by the surface AFBA, it takes in portions

of the wave front lying nearer the edge from which
momentum has been extracted to form the shear wave.

Varying the length of the boss also influences
the momentum transferred to the pellet. As the
length of the boss is increased, the region bounded
by AB and ABFA flattens so that for a pellet of
fixed thickness and fixed diameter, the surface ABFA
will at some critical boss length begin to enter the
pellet and reduce its throw-off velocity. A plot of
velocity of pellet versus boss length appears very
similar to the plot of pellet velocity versus diameter of pellet drawn in Fig. 10-10.

Finally, if the thickness of the pellet is
increased, with boss length and pellet diameter held
fixed, pellets of increasing thickness will success-
ively sense and trap momentum first from a region
ABFA which is unaffected by the surface of the boss,
then from the region AFBHA which is undergoing
expansion and in general contains less momentum per
unit volume than the region ABFA, and finally from
the region beyond that bounded by the strong shear
wave. Since much of the momentum in the shear wave
is in the forward direction, it will impart new
momentum to the relatively long pellets causing
them to fly off with greater velocity than pellets
of intermediate length, as illustrated in Fig. 10-11,
a plot of pellet velocity versus pellet thickness.
In this plot, the point A marks the pellet and B
the point at which the shear wave begins to enter
the pellet. Again simple geometric construction can
be used to establish in a semiquantitative way the
critical points along the velocity curve.

PARTICLE VELOCITY EXPERIMENTS

Pellets provide a very simple and extremely
effective tool for determining the distribution of
particle velocity and hence stress in transient
waves generated by explosions and impacts. One
technique is illustrated in Fig. 10-12. A small
cylindrical explosive charge is abutted against a
plate on the opposite side of which is affixed a
thin cylindrical pellet. When the explosive is
detonated, a transient stress wave is generated
which passes through the plate and part of its momentum is trapped by the pellet which flies off. The
velocity of the pellet can be readily measured by

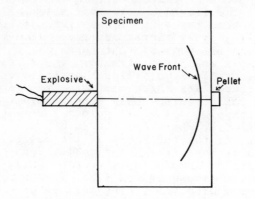

Fig. 10-12.   Experimental arrangement for determining stress distribution using
pellet technique.

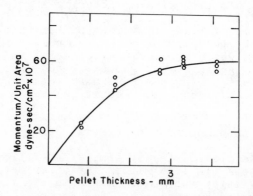

Fig. 10-13.   Momentum per unit area versus thickness
of pellet obtained by employing electric
detonator in experimental arrangement
of Fig. 10-12.

any one of a number of techniques, usually photographic, and its momentum can then be computed
from its mass. By running a series of identical
tests in which the thickness of the pellet is varied,
a momentum trapped versus pellet thickness curve of
the type shown in Fig. 10-13 can be developed. All
four of the relationships, stress-distance, stress-
time, particle velocity-distance, and particle ve-
locity-time curves for the disturbance can then be
derived from the momentum curve provided elastic
behavior is assumed.

     The particular momentum curve shown in Fig.
10-13 was obtained by exploding a small commercial
electric detonator on the surface of a 5 cm thick
plexiglas plate. The pellets were 0.63 mm diameter
cylinders ranging in thickness from 0.08 to 0.40 mm
in 0.08 mm increments. The plexiglas, which was
assumed to behave elastically, had a density of 1.18
g/cc and a dilatation wave speed of 2760 m/sec. A
stress versus distance curve, derived from Fig. 10-13
is shown in Fig. 10-14, and a stress versus time
curve in Fig. 10-15. The pulse had a maximum stress

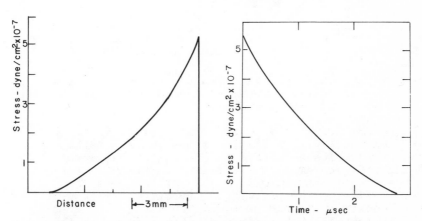

Fig. 10-14.   Stress versus distance curve derived
              from experimental data of Fig. 10-13.
Fig. 10-15.   Stress versus time curve derived from
              experimental data of Fig. 10-13.

level of about 550 atmos and a duration of about 2
microseconds. By varying the thickness of the plate
and repeating the series of experiments, a good deal
of information can be obtained as to how the wave
decays in intensity as it moves through the plate.
     With the arrangement shown in Fig. 10-16, a
single shot can be made to yield data sufficient for
the construction of a complete stress or particle
velocity distribution. This will be true provided

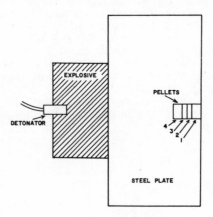

Fig. 10-16.   Experimental arrangement used to obtain
              stress transient from a single shot
              using several pellets.

the stress wave has a sharp front that is followed
by a monotonous decrease in stress level. Assume
that the explosive charge in Fig. 10-16 has been
detonated generating a sharp fronted monotonically
decreasing stress wave. As the wave reaches the
right side of the plate, the several pellets, which
are stacked in a hole bored in the face, will remain
in contact with one another until the front of the
wave is reflected from the right hand face of pellet
1 and returns to the interface between pellets 1 and
2. At the instant the front of the wave reaches the
interface, the average velocity of all elements of
pellet 1 will be greater than the velocity of the
right hand face of pellet 2 and pellet 1 will fly off.
In a similar way, pellet 2 will be flying off at a

somewhat lower velocity at a somewhat later time, the process continuing until all of the pellets have been thrown clear.

The velocity V_i of the ith pellet will be

$$V_i = 2\int_{x_{i-1}}^{x_i} v(x)\,dx/(x_i - x_{i-1}) \qquad (10.16)$$

where x_i is equal to twice the total thickness of all the pellets up to and including the ith. If the respective velocities of the several pellets from a single firing are measured, a histogram can be constructed that approximates rather closely the true spatial distribution of particle velocities in the wave. The front of the stress wave is assumed to be plane and parallel to the faces of the pellets, a condition rather closely approximated by using small diameter pellets.

An experimen ally obtained histogram is shown in Fig. 10-17. A 15.2 cm diameter, 7.6 cm thick plate was used in conjunction with a 7.6 cm diameter 5.1 cm long cylinder of explosive. The hole in the rear face of the plate was 1.3 cm in diameter and about 2 cm deep. The pellets consisted of a group of five having respective thicknesses of 0.555, 0.020, 0.24, 0.278, and 0.798 cm. The velocity of each pellet was measured close to the plate over about a 1 m interval by means of special high speed multi flash photography.

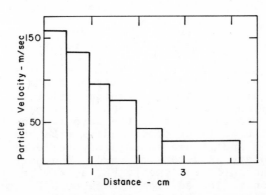

Fig. 10-17. Experimentally obtained stress-distance histogram using experimental arrangement of Fig. 10-16.

ENGRAVEMENT

When a fairly strong transient stress wave
interacts with a free surface upon which a pellet is
resting, the pellet, when it is thrown off, leaves
behind an indentation or engravement in the free
surface. The engravement has the same diameter as
the pellet, is flat bottomed, and its depth, generally a few tenths of a millimeter, depends upon the
duration and intensity of the transient stress wave.
Such engravements can be used to measure high intensity transient stresses.

To understand the dynamics of formation of this
engravement, consider the situation shown in Fig.
10-18a. The sharp front, MN of a transient stress
wave, taken in this discussion to have the form
shown in Fig. 10-19, is moving toward the right
through a semi-infinite body. The front of the wave

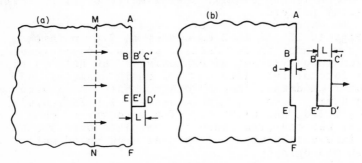

Fig. 10-18. Dynamics of formation of engravement as
             a result of impingement of stress
             transient against a free surface on
             which a small pellet rests.

is assumed plane and parallel to the free surface,
ABEF. A pellet, B'C'D'E', rests on the surface. The
situation at a somewhat later time,after the pellet
has been flung from the surface, is shown in Fig.
10-18b. When the front of the wave reaches the free
surface, the portions of it that strike the surface
elements AB and EF are immediately reflected back
into the plate as a tension wave. The portion that
strikes the surface element, BE, enters the pellet
and is soon reflected from the face, C'D', of the
pellet. Finally, when the front of the reflected
wave arrives at the interface between the pellet face,
B'E', and the surface element, BE, the pellet will

separate from the plate and begin to move off toward
the right. During the time elapsing between the
situation in Fig. 10-18a and 10-18b, the pellet will
have left a permanent impression on the surface of
the plate.

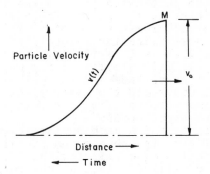

Fig. 10-19.   Profile of the stress (or particle
              velocity) transient considered in
              Fig. 10-18.

The depth of the permanent impression or
engravement can be shown to depend upon the thickness

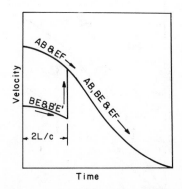

Fig. 10-20.   Velocities of the surface elements shown
              in Fig. 10-18 during and immediately
              after engravement.

of the pellet, the velocity of propagation of the

wave, and the distribution of particle velocity
within the wave. The velocities acquired by the
several surface elements involved are shown schematically as a function of time in Fig. 10-20. It is
to be remembered that at a free surface, boundary
conditions require that the particle velocity at the
free surface be just twice that of the incident
wave at any point on the wave. Therefore, during
the time interval from t = 0 to t = $2L/c_1$, AB and
EF will have just twice the velocity of BE and B'E'.
When t = $2L/c_1$, where L is the thickness of the
pellet and c_1 is the velocity of the wave in it, the
pellet leaves the surface and thereafter all surface
elements will move with the same velocity. It is
during the time interval from t = 0 to t = $2L/c_1$
that the engravement occurs.

    The distance which each surface element moves as
a function of time is shown schematically in Fig.
10-21. Neglecting edge effects, the depth of the
engravement, d, will be given by the equation

$$d = \int_0^{2L/c_1} v(t)\,dt \qquad (10.17)$$

where v(t) is the temporal distribution of particle
velocity within the wave.

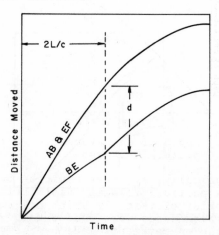

Fig. 10-21. Distances moved by the surface elements
            shown in Fig. 10-18 during and immedi-
            ately after engravement.

It is evident from this equation that the depth of the engravement will be directly proportional to particle velocity averaged from the front of the wave to a point 2L behind the front. The proportionality constant is $1/c_1$. There will be an elastically recoverable displacement but usually this is negligible but so dependent on edge effects as to be indeterminate.

A differential velocity exists between the respective surface elements only for a period of time which depends upon the wavelength of the transient disturbance. When the length of the disturbance is less than twice the thickness of the pellet, the time during which engravement occurs will be less than $2L/c_1$ and the depth of the engravement will remain constant even though the pellet thickness is increased. The situation is illustrated schematically in Fig. 10-22.

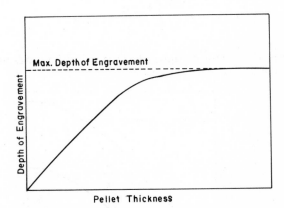

Fig. 10-22.   Depth of engravement as a function of pellet thickness.

Theoretically, all that would be necessary to construct a particle velocity distribution would be to run a series of tests in which pellets of several different thicknesses were used and to measure the depths of the engravements made by the respective pellets. The procedure used to construct the detailed particle velocity distribution would be similar to that described earlier in which pellet velocities rather than engravement depths are measured.

If both the velocity with which the pellet is thrown from the plate and the depth of the engravement are measured, then the velocity of the wave can be computed. Since the velocity V_p, of the pellet will be given by

$$V_p = 2 \int_0^{2L/c} v(t)dt/(2L/c_1) \qquad (10.18)$$

the velocity of propagation of the wave will be given by

$$c_1 = (L/d)V_p. \qquad (10.19)$$

Chapter 11

DEVELOPMENT OF NONPLANAR WAVE FRONTS AT BOUNDARIES

INTRODUCTION

The interaction of a wave with a boundary or interface generates new waves which then begin their own peregrinations through the bodies involved. These new waves often interfere with one another or with portions of the parent wave that spawned them. As a result, intense highly localized stresses tend to develop and when these are tensile, fracturing frequently occurs. A number of specific cases of such fracturing are discussed in Chapters 12 and 13.

In order to locate the regions of high stress concentration, it is necessary to establish the various wave patterns in both space and time and to fix their stress intensities. The principles of ray and physical optics, Snell's law, Huygen's principle and Fermat's principle are helpful in describing the progress and configurations of the wave fronts. The equations in Chapter 7 dealing with reflection coefficients are useful in stress levels of the individual wave front and finally the principle of superposition, full discussed in Chapter 6, provide the means for finding the stress levels developed when the waves interfere.

Although sometimes analytical solutions can be written down, more frequently geometrical construction using rule and compass is the best and often the only approach. This chapter describes how to make these constructions and applies the technique to a few illustrative situations.

157

GEOMETRICAL CONSTRUCTION AND ANALYTICAL
EXPRESSION

Generally in reflections of waves from boundaries, both dilatation and shear waves develop, the
shear wave traveling with about half the velocity of
the dilatation wave. In order to trace the progress
of the wave fronts, two basic pieces of information
are needed for each point on the front. One is the
direction it moves, usually indicated by a ray,
normal to the wave front at the point under observation. The other is the velocity with which the
front moves along the ray. The position and configuration of the wave front constantly changes with
time. In constructing wave fronts, it is helpful
and essential to remember that every point on the
wave front represents an equal amount of travel time
from its conjugate point on a previous wave front.
     The application of ray theory, adopted from
optics, is convenient to use in constructing a
series of progressing wave fronts. The problem is
to determine first the direction of the ray and
second the position of the wave front along it. The
direction is determined by assuming that the wave
front is made up of an infinite number of infinitesimally small planar elements. Each element is then
assumed to obey the laws, discussed in Chapter 7,
which control the reflection of a plane wave at a
boundary. In the case of reflection at a free surface, the new dilatation wave will go off at the
same angle, α, at which it arrives and the shear
wave will go off at an angle β given by

$$\sin \alpha / \sin \beta = c_1/c_2. \qquad (11.1)$$

Each ray thus bifurcates at the boundary, one ray of
the pair of new rays being associated with the shear
wave front and the other with the new dilatation
wave front. A point on the new dilatation wave
front progresses in the direction of its ray at velocity c_1, while a point on the new shear wave progresses in the direction of its ray at the much lower
velocity, c_2. The separation of the two fronts
increases with time. A common thing is to draw a
series of sketches each showing the positions of the
respective wave fronts at a particular instant of
time.

In constructing the diagram for a specific case,
it is useful to think in terms of sets of three rays
corresponding to the approaching wave, the reflected
dilatation wave, and the newly generated shear wave.
The progress of the respective wave fronts is
traced along these rays. Fig. 11-1 illustrates a

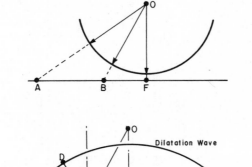

Fig. 11-1. Reflection of a spherical dilatation
            wave from a free surface.

spherical wave that is reflected from a free plane
surface. It is apparent that the time to travel
from O to B to D as a dilatation wave is the same as
to travel first from O to B as a dilatation wave
and then from B to C as a shear wave.
     The easiest way to locate, or construct, the
reflected dilatation wave front is to establish an
image point that lies below and normal to the surface
at the same distance that the source of the transient
disturbance, O, is above the surface. The reflected
dilatation wave fronts at successive times after
reflection will be concentric circles with the
image point as their centers.
     The reflected shear wave front cannot be con-
structed so easily. It will not be a spherical
surface, the radius of curvature changing from point
to point along it. The simplest procedure is to
work with several rays and compute how far the wave
has progressed along each one. The directions are

determined from Snell's law, and the distance from the relationship

$$\overline{BC} = (c_1/c_2)\overline{BD}. \qquad (11.2)$$

When the position of the wave front has been found for a number of rays, the total wave front can then be sketched.

The intensity of stress will vary along the wave front because of the change in obliquity as successive elements of the wave strike the surface. The equations needed to calculate the relative intensities of three waves that are involved were given in Chapter 7. In addition, change in stress intensity due to divergence must be taken into account, relatively easy for the spherical dilatation waves but much more difficult for the shear waves.

In drawing diagrams, it is sometimes useful to indicate the variation of intensity by variations in the width of the line showing the position of the wave front. This is done in Fig. 11-2 for a spherical wave striking the free surface of a material

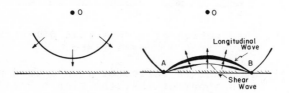

Fig. 11-2. Reflection of a spherical dilatation wave from a free surface. Widths of lines indicate distribution of energy along wave fronts.

having a Poisson's ratio of about 0.25. Points A and B indicate intersection of the incident wave with the free surface. The velocity with which these points move is known in seismology as the phase velocity. It is obvious that the phase velocity will initially be infinite at a point directly above the origin of the disturbance and then decrease monotonically, eventually reaching the velocity of the incident wave.

When making machine computer calculations, it is

necessary to express the positions of the respective wave fronts in analytical form. Taking the origin of the wave at point (0,Y) in Fig. 11-3, gives for the equation of the original wave front

$$x^2 + (y - Y)^2 = (c_1 t)^2 \qquad (11.3)$$

with t = 0 at x = 0 and y = Y where the point (0,Y) is the source of the disturbance.

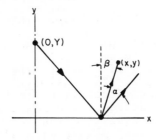

Fig. 11-3. Ray diagram used in making computer calculations of oblique reflections.

The equation for the reflected dilatation wave is

$$x^2 + (y + Y)^2 = (c_1 t)^2. \qquad (11.4)$$

The shear wave equation is best expressed in parametric form. The time t to reach the point (x,y) on the shear wave front is given by

$$t = t_1 + t_2 \qquad (11.5)$$

where t_1 is the time the wave travels as a dilatation wave, L_1/c_1 (Fig. 11-3), and t_2 is the time the wave travels as a shear wave, L_2/c_2. Thus, in terms of y

$$t = Y/c_1 \cos \alpha + y/c_2 \cos \beta \qquad (11.6)$$

and in terms of x

$$t = Y[1/c_1 \cos \alpha - \tan \alpha/c_2 \sin \beta] + x/c_2 \sin \beta. \qquad (11.7)$$

Solving explicitly for x and y yields

$$x = c_2 t \sin \beta - Y\left[(c_2/c_1)(\sin \beta/\cos \alpha) - \tan \alpha\right] \tag{11.8}$$

and

$$y = c_2 t \cos \beta - Y\left[(c_2/c_1)(\cos \beta/\cos \alpha)\right]. \tag{11.9}$$

SPHERICAL WAVE STRIKING A PLANE BOUNDARY

The interaction of an obliquely incident plane wave with loose or cohesive plane boundaries, and between similar and dissimilar materials was discussed in Chapter 7. In considering the interaction of a spherical wave front with these types of plane boundaries, the incident wave front can again be treated as if it consists of an infinite number of infinitesimally small planar elements. In general, there are five wave fronts: the incident wave, the shear and dilatation reflected waves, and the shear and dilatation transmitted waves. Again each wave front is an equal time path from the source.

Two situations must be distinguished, one in which $c_1 > c_1'$, and the other in which $c_1 < c_1'$. The case of $c_1 > c_1'$ is illustrated in Fig. 11-4. In order not to complicate the drawing too much only the dilatation waves are shown. The reflected wave

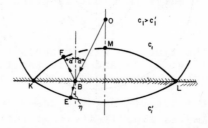

Fig. 11-4. Reflected and transmitted dilatation
         wave fronts generated during inter-
         action of a spherical dilatation wave
         with a cohesive interface between
         two dissimilar materials. $c_1 > c_1'$.

front KML has the same radius of curvature as the
incident wave front and can be easily located using
an image source. The curvature of the transmitted
wave front is not constant from point to point along
it; it becomes flatter as the distance between K and
L increases. Points on the transmitted wave front
can be located by noting that

$$\overline{OB}/c_1' + \overline{BE}/c_1 = (\overline{OB} + \overline{BF})/c_1 \qquad (11.10)$$

and remembering that

$$\sin \alpha / \sin \eta = c_1/c_1'. \qquad (11.11)$$

The second situation, $c_1 < c_1'$, is illustrated in
Fig. 11-5 for the case where $2c_1 = c_1'$. The character of the transmission changes completely at
point A when

$$\alpha = \sin^{-1} (c_1/c_1'). \qquad (11.12)$$

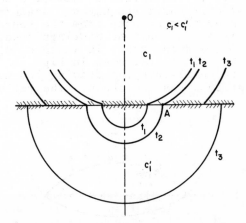

Fig. 11-5. Same as Fig. 11-4 except incident wave
front and only transmitted dilatation
wave front shown. $c_1 < c_1'$. t_1, t_2,
and t_3 indicate the situation existing
at three successive times.

For angles equal to or greater than this, the wave
will be totally reflected. Note that this critical
condition arises at just the time when the velocity

with which the incident wave cuts the boundary becomes equal to $c_1^\dagger$.   For larger values of α, the
upper wave will sweep across the boundary at a
velocity less than $c_1^\dagger$.   Consequently the lower wave
will outrun the upper and in so doing will induce a
new wave back into the upper layer. This pattern of
behavior forms the basis of extremely useful geophysical refraction methods for locating underlying
high velocity geological structures.

Relative stress intensities along wave fronts
are differentiated graphically by line thicknesses
(see Fig. 11-2).   This is done in Fig. 11-6 for the
case of a loose boundary separating two blocks of
the same material.   Here the curvatures of the three
dilatation waves are all the same. Curvatures of
both the shear waves are identical to each other and
the same as the curvature of the shear wave drawn in
Fig. 11-2.

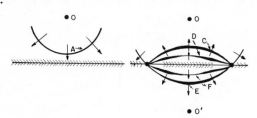

Fig. 11-6. Positions and relative intensities
            of shear and dilatation waves gen-
            erated during reflection of a spherical
            dilatation wave at a loose boundary.

REFLECTION OF A PLANE WAVE FROM A CURVED FREE
SURFACE

Reflection of a plane wave at a curved boundary
is treated as a series of interactions between
successive points along the wave front with infinitesimally small planar elements, each tangent to the
boundary at one point of intersection of the wave
front and the boundary. The positions of the reflected wave fronts are found in the same way as in
the preceding sections of this chapter. The partitioning of energy between shear and dilatation
waves, depending as it does upon angle of incidence,
will vary from point to point along the wave front.

In general, a convex free surface transforms a plane wave into a converging wave just as an optical lens or mirror focuses light rays. If the incident wave is compressive, the converging wave will be tensile so that at its focus, enormous tensile stresses are built up which often produce local fracturing.

POINT LOADING OF A CYLINDER OR SPHERE

A good illustration of reflection from a curved surface is that in which an impulsive point load is applied to the surface of a cylinder or sphere as indicated in the first drawing of Fig. 11-7. The configuration of the reflected dilatation wave is shown at four successive times. The reflected shear

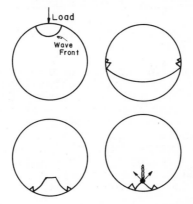

Fig. 11-7. Stages in wave front development resulting from application of a point load to the surface of a cylinder or sphere.

waves are not shown. As the original divergent dilatation wave moves down, it is being continually reflected, resulting in the development of several cusps in the reflected wave front with very intense stress concentrations developing within the cusps, the most important of which is near the opposite side of the cylinder. It is not at all unusual for local fracturing to develop in this region.

REFLECTION OF PLANE WAVE FROM CIRCULAR OPENING

An excellent representative example of how highly localized stress concentrations can evolve from wave reflection is when a plane impulsive wave reflects from two circular cylindrical openings.

This situation is illustrated in Fig. 11-8. An isotropic medium contains two infinitely long, parallel, cylindrical holes of equal radius. A longitudinal compression wave generated by an impulsive

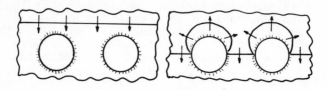

Fig. 11-8. Reflection of a plane wave from two circular openings.

load propagates through the medium with its wave front parallel to the plane which contains the axes of the holes. In general, the wave front impinges obliquely upon the free surfaces of the holes, setting up two new waves, a reflected longitudinal wave and a generated shear wave. Only the development of the reflected longitudinal waves is illustrated in the figure, since the shear wave will lag behind the longitudinal wave fronts. The two reflected longitudinal waves, which will usually be primarily tensile, will meet along a plane which lies halfway between the hole axes and is perpendicular to the plane containing them. The tensile stress resulting from superposition of the two waves can be quite high, so high, in fact, that fracturing frequently develops along this plane in specimens subjected to this type of loading.

The location of the reflected longitudinal wave front as a function of time can be written as a pair of parametric equations. Consider what happens around the circular hole shown in Fig. 11-9. A ray, A'B, which strikes at B making an angle α with the normal to the free surface of the opening, will change its direction by the angle $(\pi - 2\alpha)$, advancing then in the direction BC. The point C, consid-

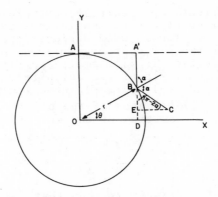

Fig. 11-9. Diagram used for developing parametric
          equations describing position of wave
          front generated by reflection of plane
          wave from circular opening.

ered to be the location of the total advance of the
ray, will lie on the new wave front. Now taking
t = 0 at the instant the wave strikes the opening
at the point A, that point on the wave front corresponding to the deviated ray A'BC will have required
the time t to travel the distance A'B + BC, which
can be written

$$t = t_{A'B} + t_{BC}. \qquad (11.13)$$

where $t_{A'B}$ is the travel time from A' to B and t_{BC}
is the travel time from B to C.
    Inspection of Fig. 11-9 yields

$$t_{BC} = t - (r - r\sin\theta)/c_1 \qquad (11.14)$$

which can be rearranged, giving

$$t_{BC} = \left[tc_1 - r(1 - \sin\theta)\right]/c_1 \qquad (11.15)$$

and since

$$\overline{BC} = t_{BC}c_1 \qquad (11.16)$$

substitution gives

$$\overline{BC} = tc_1 - r(1 - \sin \theta) \qquad (11.17)$$

where r is the radius of the cylindrical hole, c is the velocity of propagation of the dilatation wave, and θ is the angle between the x axis (Fig. 11-9) and the radius vector to the point of reflection from the axis of the perimeter of the hole.

The x, y coordinates of the point C, which establishes the position of the wave front are

$$x = \overline{OD} + \overline{EC} \qquad (11.18)$$

$$y = \overline{BD} - \overline{BE} \qquad (11.19)$$

where from Fig. 11-9 it is seen that

$$\overline{OD} = r\cos \theta; \quad \alpha = 90 - \theta \qquad (11.20)$$

$$\overline{EC} = \overline{BC}\sin(\pi - 2\alpha) = \overline{BC}\sin 2\theta \qquad (11.21)$$

which on substitution gives

$$\overline{EC} = \left[c_1 t - r(1 - \sin \theta) \right] \sin 2\theta \qquad (11.22)$$

and

$$\overline{BD} = r\sin \theta \qquad (11.23)$$

$$\overline{BE} = \overline{BC}\cos (\pi - 2\alpha) = \overline{BC}\cos 2\theta \qquad (11.24)$$

which on substitution gives

$$\overline{BE} = \left[c_1 t - r(1 - \sin \theta) \right] \cos 2\theta. \qquad (11.25)$$

Substituting these values now into Eqs. (11.18) and (11.19) gives the parametric equations of the reflected wave front

$$x = r\cos \theta + \left[c_1 t - r(1 - \sin \theta) \right] \sin 2\theta \qquad (11.26)$$

$$y = r\sin \theta - \left[c_1 t - r(1 - \sin \theta) \right] \cos 2\theta \qquad (11.27)$$

which can be easily plotted for various values of t. These equations are also useful in making computer calculations.

DIFFRACTION OF A COMPRESSIVE WAVE NEAR A NOTCH

Encounter of a simple transient compression wave with the boundaries of a re-entrant corner or notch will cause a whole new and complicated pattern of wavelets and waves to develop. Near the corner most of the energy of the original wave is transformed into additional shear and dilatation waves, resulting in diffraction of energy around the notch. The phenomenon corresponds in some respects to the diffraction of light around a corner.

Specifically, consider the situation shown in Fig. 11-10. A sharp fronted compression wave orig-

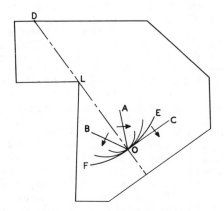

Fig. 11-10. Wave pattern generated by diffraction
of a plane wave at a notch.

inates at point D.   The wave encounters the notch L where the front of the wave is chopped in two, part (not shown) of the front being reflected and the remainder moving on forward unobstructed.   The forward advancing part of the wave front, OC, begins at once to generate new shear waves and dilatation waves. By feeding its own original energy into the new waves, it thereby weakens itself. The cutting of the wave front at the notch L leaves a highly stressed free end, O, through which energy of the wave will escape to the left into the unstressed material at its side. As a consequence there develops around O the pattern of waves and wavelets drawn in the figure, the situation being somewhat analagous

to that of a wave moving along a free boundary discussed in Chapter 8. The two principal new waves
are shear waves: one, OB, entering the unstressed
region to the left; and the other, OA, moving into
the stressed region behind the wave front. Both
wave fronts are inclined to the original wave front,
OC. The sine of the angle between each shear wave
front and the normal to the wave front OL is equal
to the ratio of the transverse or shear wave velocity to the dilatation wave velocity.

Two other regions of varying state of stress
develop also: one between the arc OF and the line
OB, and the other between the arc OE and the line
OA. The curve EOF is the arc of a circle having its
center at the point L, the tip of the corner.
Neither OF nor OE is a coherent wave front since the
source O, feeding energy to the regions AOE and BOF,
is moving with the wave front OC, making it impossible to construct an envelope to the individual
wavelets. The arcs, OF and OE, simply represent
the limits to which the influence of the terminal
point O can be felt. The region, FOB, lying to the
left of O, contains compression strain energy, the
energy having been introduced when the transient
wave by its passage suddenly juxtaposes a highly
stressed region beside the unstressed region FOB.
The region EOA, to the right of O, is the region
from which strain energy has been extracted to
supply energy to the regions FOB, LOB, and LOA. In
the two regions, FOB and EOA, it is meaningful to
describe the transient state in terms of energy or
momentum density rather than wave fronts, a crude
measure of the density being the spatial concentration of wavelets. During passage of the original
wave, the intensity of its front OC degrades in the
neighborhood of the point O, resulting in the development of a front along which the intensity is no
longer uniform. While a complete and rigorous
analytical description of the process is exceedingly
difficult, the foregoing pattern of waves is qualitatively realistic and representative of many other
similar situations.

CHAPTER 12

STRESSES DEVELOPED IN A CORNER

INTRODUCTION

A fracture surface is often generated in the corner formed by the intersection of two free surfaces when a transient compression wave enters it. The fracture takes place as a result of a highly localized concentration of stress that develops as a consequence of reflection of the stress wave from the two sides of the corner.

As a simple example, consider the situation shown in Fig. 12-1. A compressive longitudinal wave enters a 90° corner where tensile waves are generated as it reflects from the two sides of the corner. Soon the fronts of the two tensile waves will collide head on, encounter taking place simultaneously along the line AF, where a strong, highly localized tensile stress is set up. Under such conditions fracturing commonly occurs.

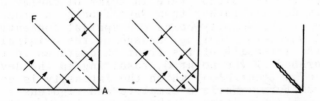

Fig. 12-1. Mechanics of the generation of a corner fracture.

Usually the situation is more complicated than this where either the corner is not at right angles, the wave front is not plane, or the stress is of

finite duration, inhibiting the recycling of the
compressed material to a state of tension. Sometimes
also the shear waves, which are neglected here,
generated by the reflections, play an important role
leading to fracturing when they collide with other
shear or longitudinal waves. The stresses which are
responsible for causing corner fractures can be
readily determined by applying the laws of reflection
given in Chapter 7 in conjunction with the principle
of superposition discussed extensively in Chapter 6.

BASIC GEOMETRY

     The nature and salient aspects of the inter-
actions that take place among the waves generated in
a corner are perhaps best illustrated by considering
several representative specific examples: an acute
angle, a right angle, and an obtuse angle. Special
situations, which will be discussed later, develop
when the wave enters the corner angle with its front
oriented normal to the bisector of the angle.
     The acute angle case is depicted in Fig. 12-2.

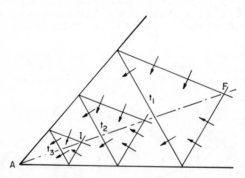

Fig. 12-2.   Reflections associated with entrance of
             plane wave into acute angle corner.
             t_1, t_2, and t_3 indicate wave patterns
             at successive times.

The front of the incident plane wave and its associated reflections are shown for three positions, t_1,
t_2, and t_3, as the wave progresses down into the
corner. The shear wave fronts are not shown. As the

wave moves, the two reflected dilatation wave fronts
begin to cross over one another; the trajectory of
the point of intersection is along the straight
line AF. The distance between the point of intersection and the original wave front grows less and
less as the apex of the angle is approached, becoming
zero when it is reached.

The relationships existing among angles are
shown for this example in Fig. 12-3. The wave front,

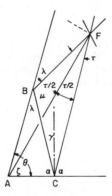

Fig. 12-3.   Angles associated with reflection of plane
              wave into acute angle corner. BC is
              incident wave front. BF and CF are
              reflected wave fronts.

inclined at an angle α with respect to the lower side
of the corner, is assumed to enter a corner having an
angular spread of θ. The line AF along which the
reflected waves begin their intersection is inclined
at an angle ζ with the bottom of the corner, given by

$$\zeta = \theta - (90 - \alpha). \qquad (12.1)$$

The angle of intersection, τ, between the two reflected wave fronts is given by

$$\tau = \alpha + \lambda - 180 \qquad (12.2)$$

where λ is the angle lying between the incident wave
front and the side of the corner. The line of inter-

section AF bisects the angle τ.

It should be noted that the maximum value that either α or λ can assume is 90º if reflection is to take place.

For the very special case where the wave front is normal to the bisector of the corner angle θ

$$\zeta = \theta/2, \qquad (12.3)$$

the intersection proceding along the bisector.

Similar diagrams are drawn in Figs. 12-4 and 12-5 for a right angle corner. The situation is quite different from that of an acute angle. Intersection of the two reflected waves consists of a head on, instantaneous collision along the line AF (Fig. 12-4) that passes through the apex of the corner and is inclined with the bottom edge of the corner at the angle ζ which is equal to the angle of incidence γ that the incoming wave makes with the bottom edge.

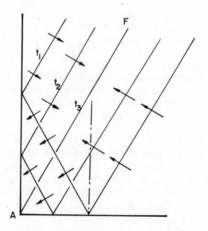

Fig. 12-4.   Reflections associated with entrance of plane wave into right angle corner. t_1, t_2, and t_3 indicate wave patterns at successive times.

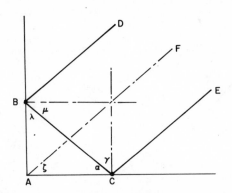

Fig. 12-5. Angles associated with reflection of
            plane wave into right angle corner.
            BC is incident wave front. BD and CE are
            reflected wave fronts.

     The order of events for an obtuse angle corner,
diagrammed in Fig. 12-6, is still different. Intersection of the reflected waves begins only after the
incident wave has reached the apex. The point of
intersection then moves upward back into the material.
The angles that become involved are identified in
Fig. 12-7. The angle μ that the line of intersection
forms with the bottom side of the corner will be
given by

$$\mu = \theta - \gamma \qquad\qquad (12.4)$$

where γ is the angle of incidence of the wave on the
bottom surface. The important limiting case,
θ = 180°, of an obtuse angle corner was discussed in
detail in Chapter 7.
     Similar to the intersection in an acute angle
corner, the reflected portions of a wave entering an
obtuse angle corner will proceed along the bisector,
but in this case away from the apex instead of toward
it.

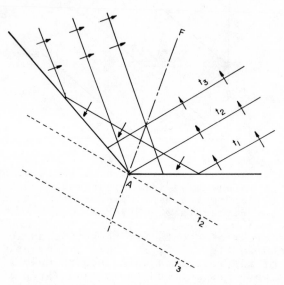

Fig. 12-6.   Reflections associated with entrance of
            plane wave into obtuse angle corner.
            t_1, t_2, and t_3 indicate wave patterns
            at successive times. Dashed lines
            indicate positions wave front would have
            assumed if corner was not present.

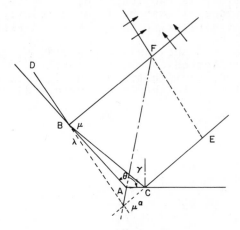

Fig. 12-7.   Angles associated with reflection of
             plane wave into obtuse angle corner.
             BC is incident wave front.   BD and CE
             are reflected wave fronts.

COMPUTATION OF PRINCIPAL STRESS

     Thus far the duration of the wave has been
assumed to be very short.   In fact, the diagrams of
the previous section were drawn as if the wave was
infinitely thin, consisting only of a front.   Transient waves, in general, of course, are of finite
duration making it necessary to take into account
progressive reflection of succeeding parts of the
wave.   When this is done, it is convenient in making
computations to distinguish between two types of
situations: one, in which the interference between
the two reflected waves takes place near the apex
where the material is still being compressed by the
incident wave; and the other, where interference
takes place relatively far from the apex, outside the
region being influenced by the incident wave.
     For each pair of wave entrance and corner angles
there exists a point along the line of intersection
of the reflected waves which separates the close in
region from the far out region.   In order to locate
this point, the length of the wave must be specified.

For a 90° corner angle, the point lies $(1/2)\Lambda$ from
the apex where Λ is the length of the wave. For
other situations, it can be located readily by making
constructions similar to those shown in Figs. 12-2,
12-4, and 12-6.
    Far from the apex only two stresses, correspond-
ing to the reflected waves, are involved and these
are superposed, as illustrated in Figs. 12-8, 12-9,
12-10, and 12-11, to obtain the principal stress.

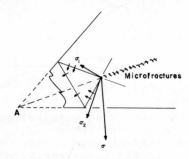

Fig. 12-8.   Stress generated by superposition of
              reflected waves. Acute angle; far from
              apex A.

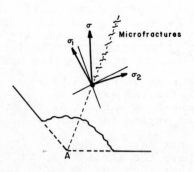

Fig. 12-9.   Stress generated by superposition of
              reflected waves. Obtuse angle; far from
              apex A.

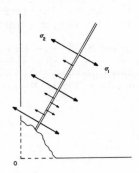

Fig. 12-10.   Stress generated by superposition of
              reflected waves.   Right angle; far from
              apex A.

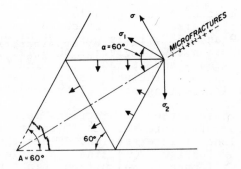

Fig. 12-11.   Stress generated by superposition of
              reflected waves. 60° angle; special case
              where wave front is normal to bisector
              of angle; far from apex A.

   The simplest case is shown in Fig. 12-10 for a
90° corner. Here the stresses operate in the same
direction so that they can be added algebraically.
The relative stress intensities of one reflected
wave, σ_1, and the other reflected wave, σ_2, will
depend upon the orientation of the wave front with
respect to the sides of the corner and can be calcu-

lated using the reflection coefficient equation given
in Chapter 7.

Computation of the principal stresses for an
acute angle case as in Fig. 12-8 and for an obtuse
angle case as in Fig. 12-9, is more complicated,
involving not only the application of the reflection
coefficient equations to calculate intensities, but
also use of the superposition equations, Eqs (6.19)
and (6.20), to find the magnitude and direction of
the principal stress. Generally, here the resultant
principal stresses will be neither parallel nor
perpendicular to the line of intersection.

A special case exists when the wave enters the
corner along the bisector (Fig. 12-11). Now σ_1 and
σ_2 are equal since the wave front is inclined at the
same angle to both sides of the corner. The orientation, ϕ, of the principal stress with the line of
intersection of the waves is determined by the
equation

$$\tan 2\phi = \left[(\sigma_2 - \sigma_1)/(\sigma_2 + \sigma_1)\right] \tan 2\alpha = 0 \qquad (12.5)$$

so that ϕ must be either parallel or perpendicular to
the line of intersection. For corner angles lying
between 45° and 135°, it is perpendicular; for corner
angles less than 45° or greater than 135°, it is
parallel. At 45° and at 135°, ϕ is indeterminate,
all directions of principal stress are equally likely,
and a most interesting state of hydrostatic tension
must exist.

When these latter interactions lead to fracturing, the macrofracture surface is constrained to lie
along the line of intersection. Since the intersecting waves are obliquely inclined, a series of microfractures will open up which then join together to
form the final fracture surface. Each microfracture
will lie perpendicular to the normal tensile stress
that caused it. The orientations and locations of
these microfractures have been indicated on Figs.
12-8, 12-9, 12-10, and 12-11.

In all cases, close to the apex, three stresses
are involved: the two reflected stresses and the
stress in the incident wave. The resultant principal stress, which depends strongly on its location
along the line of intersection, is calculated by
superposing two of the stresses and then combining
the resultant with the third. The specific case of
a sawtooth wave is discussed in the next section.

SOME SPECIFIC STRESS COMPUTATIONS

The development of stress within an acute angle corner due to the reflection of a sharp fronted sawtooth wave of maximum intensity σ_o and length Λ is now considered as a specific example. The situation far in from the apex is illustrated in Fig. 12-12 and close to the apex in Fig. 12-13. The coefficient

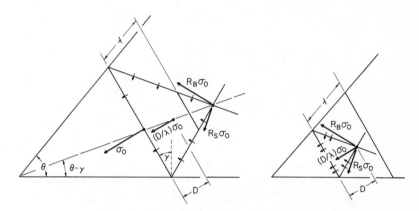

Fig. 12-12.   Development of stress within acute angle by reflection of sharp fronted sawtooth wave of wavelength Λ. Far from apex.

Fig. 12-13.   Same as Fig. 12-12 except wave has advanced so that superposition of reflected waves is occurring within region stressed by incident wave.

of reflection for portions of the wave reflected from the bottom of the corner and the side of the corner are labeled R_B and R_S, respectively. The linear variations of stress within the sawtooth wave is given by $(D/\Lambda)\sigma_o$ where D is distance measured from the front of the wave.

The calculations shown in Figs. 12-14, 12-15, 12-16, 12-17, and 12-18 were made by superposing the two stresses $R_B\sigma_o$, $R_S\sigma_o$, and $(D/\Lambda)\sigma_o$ in the region close to the apex. The reflection coefficient R is a function of both angle of obliquity of the wave and the Poisson's ratio of the material (see Fig. 7-10) and the dependence of R on both must be taken into account. The resultant stress state at each of

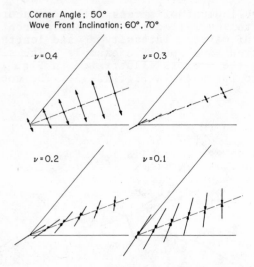

Fig. 12-14.   Stress fields developed during corner
reflection.   Arrows indicate directions
and magnitudes of stresses.

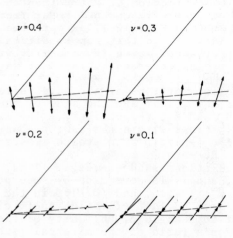

Fig. 12-15.   Same as Fig. 12-14.   Different conditions.

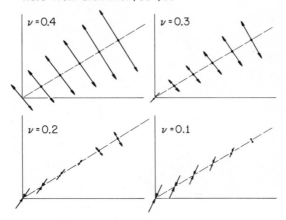

Corner Angle; 90°
Wave Front Inclination; 30°,60°

Fig. 12-16.   Same as Fig. 12-14.   Different conditions.

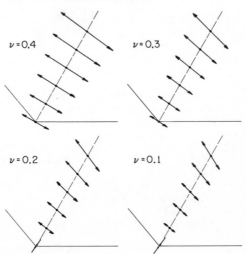

Corner Angle; 130°
Wave Front Inclination; 20°, 30°

Fig. 12-17.   Same as Fig. 12-14.   Different conditions.

Corner Angle; 150°
Wave Front Inclination; 5°; 25°

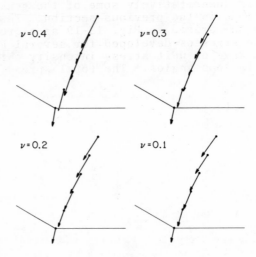

Fig. 12-18.   Same as Fig. 12-14.   Different conditions.

several points is represented either by two opposing
stress vectors or by a single vector depending upon
which representation is more convenient.   These
vectors describe the magnitude and direction of the
principal stress in each interfering wave.   For each,
there is a point to the right of which all points
lying on the line of intersection are subjected to
the same maximum principal stress.   This boundary
point is the last for which stress vectors are shown
in the drawing.

Generally for corner angles between 45° and 135°,
the resultant stress lies roughly perpendicular to
the line of intersection of the stresses.   The drawings for the acute, 50°, angle in Figs. 12-14 and
12-15 are good examples.   For obtuse, 130° and 150°,
corner angles in Figs. 12-17 and 12-18, the stress is
everywhere tensile and roughly parallel to the line
of intersection.

Since the respective orientations of the microfractures that develop are controlled by the directions of the principal stresses, in obtuse angles the

microfractures will generally be oriented perpendicular to the line of intersection.

In the region far in from the apex, the computations confirm quantitatively some of the general
statements made in the previous section. The computations are summarized in Fig. 12-19 as a group of
plots of the stresses developed for several Poisson's
ratios by a wave of unit stress intensity entering
corners of various angles. The total spread in

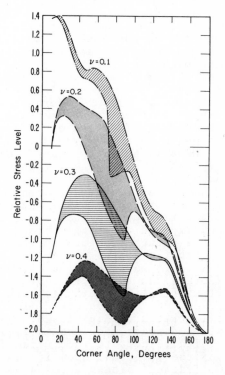

Fig. 12-19. Stress relative to incident wave generated far from apex as function of corner
angle and Poisson's ratio. Each shaded
portion delineates spread in values of
relative stress for a particular Poisson's
ratio for all permissible wave front
inclinations.

values for all permissible wave front inclinations is
confined to the region lying between the two curves
which are plotted for a single value of Poisson's
ratio. The dominant role that Poisson's ratio plays
in determining what stress will develop is clearly
evident from inspection of the curves. Relevant to
the possibility of fracturing, it is noteworthy that
for materials having low Poisson's ratios, little
tensile stress develops until the corner angle becomes
greater than about 90°. In acute corner angles, the
resultant stress is compressive and corner fracturing
would not be expected to occur.

It is also evident from Fig. 12-19 that the
relative stress is not especially sensitive to the
angle of inclination of the entering wave front. This
insensitivity could be anticipated since the entering
wave front which might be steeply inclined to one side
of the corner will form a small angle with the other
side, and vice versa.

Close to the apex, for a 90° corner as well,
(Fig. 12-16), superposition of stresses is more
involved. Both the magnitude and direction of the
resultant principal stress vary rapidly as the apex
is approached. Usually but not always, compressive
stresses develop, the exception being in materials
with high Poisson's ratios.

There is an unusual feature of corner fractures
that occurs in plastics and metals. Whereas the
inclination of the fracture to the free surface of
the corner is easily and accurately predictable, it
is observed that a corner fracture does not usually
extend clear to the tip of the corner and generally
stops short of it. It may then bend off toward one
surface or the other and run up to that surface.
Occasionally it will split and run to both surfaces.
Four commonly experimentally observed fracture
patterns are illustrated in Fig. 12-20. Which
situation develops depends upon the inclination of
the wave front with respect to the surfaces, upon
the intensity of the wave, and upon its duration.
Always when the fracture breaks through to a surface,
it does so along the shortest path. When the path
is indeterminate, as is the case of a fracture inclined at 45° to both surfaces, the breaks go to both
surfaces; if the wave is weak, to neither. The
deviatory part of the fracture is less long for
intense waves than for weak ones, the principal
corner fractures extending close to the tip of the
corner.

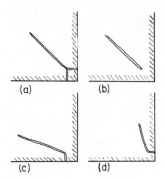

Fig. 12-20.   Commonly observed corner fracture
              patterns.

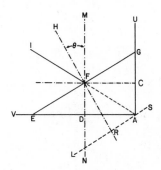

Fig. 12-21.   Interactions leading to fracture
              patterns of Fig. 12-20.

The interactions that generate such fracture
patterns are illustrated in Fig. 12-21, drawn here
for the special case of a 90° angle and for a flattopped, sharp fronted wave of short duration (compare
Fig. 12-16 for a sawtooth wave). The drawing illustrates the situation obtaining at the exact instant
that the wave fronts are combining to form the
corner fracture. The line SL is the position that
the wave front would have reached had it not been
reflected and the line GE is the position of the rear
of the wave. Thus the distance FR corresponds to the
duration of the disturbance. The wave front makes an

angle θ with the lower surface, VA, of the corner.
The fracture will develop along the line AI, where
the fronts of the two reflected waves meet, making an
angle φ with the side surface of the corner, where
φ = 90 - θ. From the point F inward toward the point
I along the fracture line AI, the fronts of the waves
are meeting and interfering in a region which is unstressed because all of the incident wave has passed
beyond it, the rear of the wave at the time of the
encounter lying along the line EG. From the point A
to the point F along the same line, the situation is
quite different, the material being under compression
immediately before the waves meet. Thus the tensile
stress, tending to generate a fracture, is much higher from the point F toward the point I than it is
from A to F. It is entirely possible that the tensile
stress along AF will be too low to generate a fracture
whereas beyond F, where the stress is higher, a
fracture will develop. In fact, the tensile stress
is very likely to be too low to generate a fracture
since the two interfering waves are very much less
intense than the incident wave because of the energy
partitioning between dilatation and shear wave occurring during the oblique reflections of the waves.
Once a fracture has taken place along the line FI,
the fracture may then extend itself to either of the
two surfaces, to both, or terminate at F.

Observations of explosively attacked specimens
confirm that the fracture patterns conform to those
drawn in Fig. 12-20. In general, experimental observations show that when the wave is strong, the fracture runs to the bottom surface of the corner when
the distance FD (Fig. 12-21) from the tip of the
corner fracture to the bottom surface is less than
the distance FC to the side surface; it runs to the
side surface when FD is greater than FC; it goes to
both surfaces when FD is approximately equal to FC.
When the wave is weak, the fracture goes to neither
surface. All of these experimental observations are
in good qualitative agreement with predictions from
the above theoretical considerations.

FRACTURES IN SOME METAL-EXPLOSIVE SYSTEMS

There are any number of metal-explosive systems
in which corner fracturing occurs. Often the results
are rather unexpected, particularly when looked at

from the point of view of static loading, though
readily explicable in terms of principles regarding
the influence of boundaries on transient stress waves.

An explosively internally loaded hollow, heavy-
walled, rught circular cylinder is one good example.
A simplified schematic of the stress situation shortly after initiation of the explosive is shown in
cross section in Fig. 12-22. When detonated in
contact with the metal, the charge initiates a sharp
fronted high intensity stress wave within the wall of
the cylinder. The loading of the cylinder is asymmetrical since the detonation proceeds through the
explosive with a finite velocity D. The induced

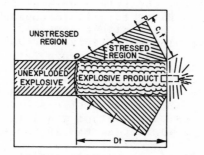

Fig. 12-22.   Stress situation shortly after initiation
              of an explosive in an internally loaded
              cylinder.   PQ is advancing conical
              compressive wave front.   Explosive has
              detonated as far as point Q.

stress wave moves both outward and downward through
the cylinder, and as it propagates there exist simultaneously in the wall of the cylinder, a very high
tensile stress σ_θ in the circumferential direction, a
compressive stress σ_r in the radial direction, and
a compressive stress σ_z parallel to the longitudinal
axis of the cylinder. The magnitudes of these three
stresses decrease rapidly due to divergence as the
wave propagates through the cylinder wall.

A circular cross section of the cylinder would
show the vector component of the stress wave in that
plane as a circularly expanding wave with the intensities of the normal and tangential stresses decreasing as the wave front expands radially outward. The

transition is abrupt between a conical portion of the cylinder which will be in a stressed condition and the remaining portion which will be at rest in an unstressed condition. The wave front PQ will be inclined at an angle β with the axis of the cylinder, where β is given by

$$\beta = \sin^{-1} (c_1/D). \tag{12.6}$$

By the time shown, the terminus Q of the wave front will have progressed a distance Dt along the cylinder wall.

When the stress wave strikes the outer free surfaces of the cylinder, it will be reflected. In examining the effects of these reflections, it is helpful to consider the stress wave as a combination of a wave of high circumferential tensile stress that advances simultaneously with a compression wave, the latter wave involving both radial and axial stresses. An examination of particle motion within the reflected stress wave indicates that the circumferential stress component remains a tensile stress and tends to reinforce the normal stress established in the cylinder wall by the incident stress wave. The compressive stress component of the incident stress wave appears as a tensile stress in the reflected wave.

The progress of the compressive stress wave and the ensuing interaction of the reflected elements of the wave are illustrated in Fig. 12-23. Observations on the angle of fracture, α, given by

$$\alpha = \cos^{-1} (c_1/D) \tag{12.7}$$

can be used to calculate either the wave velocity c_1, or the detonation velocity D, depending upon which of the two is known.

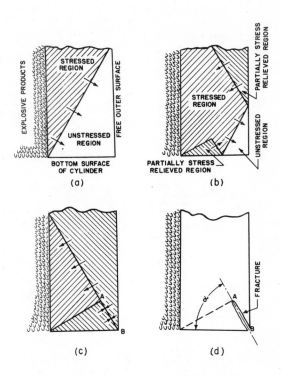

Fig. 12-23.   Propagation and interaction of stress
               waves in wall of cylinder of Fig. 12-22.

    Other examples of corner fracturing in metal
cylinders loaded internally with explosives are
shown in Figs. 12-24, 12-25, and 12-26. In Fig.
12-24, an explosive charge is detonated inside a
"square cylinder", creating the compressive, divergent, longitudinal, sharp fronted wave, the radial
component of which is drawn in Fig. 12-24a. As the
front of the wave reaches the surfaces, release or
tension waves begin to propagate back into the
compressed region. As the fronts of the release
waves meet, high tensile stresses will exist momen-
tarily along the diagonals of the cylinder section
and the cylinder breaks into four pieces along these
surfaces. Unlike its statically loaded counterpart,
the cylinder breaks along its thickest sections.

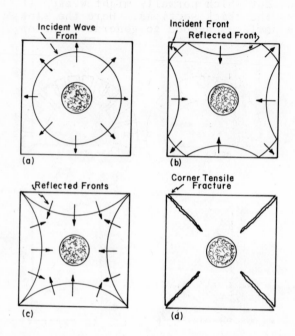

Fig. 12-24.   Interactions leading to corner fractur-
              ing in internally explosively loaded
              "square cylinder".

In Fig. 12-25, a star-shaped piece is seen to
fracture at all of its thickest sections.

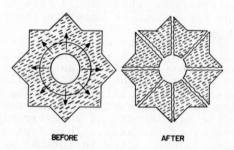

Fig. 12-25.   Fracturing of internally explosively
              loaded star-shaped cylinder.

The cylinder in Fig. 12-26, in which deep slots
have been cut which normally might weaken it, breaks
up across the thick sections. Here the slots provide
the free surfaces needed to generate tension or
release waves.

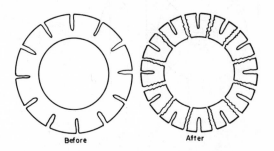

Before   After

Fig. 12-26.   Fracturing of internally explosively
              loaded slotted cylinder.

POINT AND LINE LOADING AND BREAK UP OF BLOCKS
AND CYLINDERS

The shape of the body which is explosively
loaded will exercise significant control over how it
breaks up. Illustrative and representative sets of
shapes and the fractures that might be expected when
a point impulsive load is applied are shown in
Figs. 12-27 through 12-32. Fractures arising from
causes other than reflections in corners, such as
spalling and radial fracturing, are not shown. All
shear wave interactions which would have to be superimposed on the ones shown to complete the fracture
pattern have been ignored.
Figure 12-27 shows the central fracture generated in a rectangular block when the load is applied
along the axis. This situation can be thought of as
a zero degree corner. Right circular cylinders are
especially susceptible to axial fracturing since in
such cases the reflected wave is highly convergent
and leads to the development of an extremely high
tensile stress along the axis. This type of fracturing will be explored later in further detail.

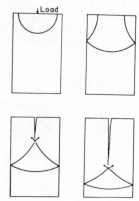

Fig. 12-27.   Interactions leading to generation of central fracturing in rectangular block of cylinder due to point loading.

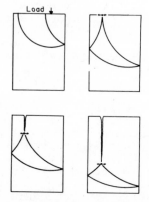

Fig. 12-28.   Similar to Fig. 12-27 except point load is applied off center.

The load applied asymmetrically to a rectangular block illustrated in Fig. 12-28 will generate a fracture but in this case the fracture lies not directly under the load but in a mirror image position. Fracturing in a single wedge, Fig. 12-29, will be less intense than in a square plate because part of the wave is lost, being unable to reflect from the sloping side. This reduction in intensity of reflected waves also occurs in the case of a double wedge, Fig. 12-30. Note that the single wedge is as effective in reducing fracturing as the double wedge,

the elimination of one of the two interfering waves
being all that is required to prevent the fractures.

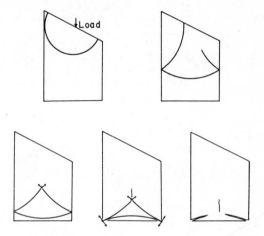

Fig. 12-29.   Fractures generated in a single wedge
              by point load.

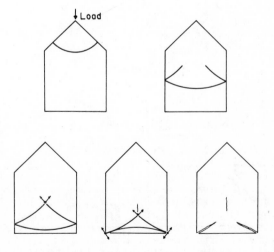

Fig. 12-30.   Fractures generated in a double wedge
              by point load.

Corner fracturing in plates of varying thicknesses is illustrated in Fig. 12-31.   As the width to
thickness ratio decreases, the corner fracture becomes
less inclined to the bottom surface.   The asymmetry of
corner fracturing produced by asymmetric loading is
illustrated in Fig. 12-32.

These several aspects of fracture can be effectively demonstrated by detonating blasting caps on
small blocks of plexiglas, in which, since it is
transparent, internal fractures can be easily and
readily seen.

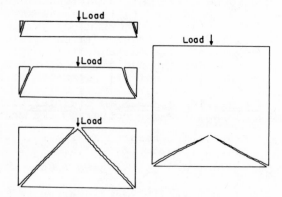

Fig. 12-31.   Corner fracturing in plates of varying
thicknesses generated by point load.

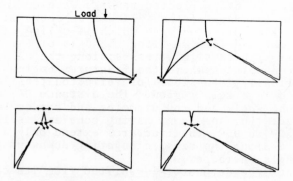

Fig. 12-32.   Asymmetrical corner fracturing generated
in plate by asymmetrical application of
point load.

FRACTURING IN END LOADED CYLINDERS

Shear wave interactions have been ignored thus far in the discussion of corner fracturing. In most cases the faster traveling dilatation waves will arrive first and effectively obscure the result of any action between shear waves or shear and dilatation waves. Knowing the values of reflection coefficients and velocities of propagation it is, of course, possible to compute stress concentrations using the superposition equations developed in Chapter 6. Although in most instances, the computations become too involved to be practical, experimental work in which explosives were used to induce shocks into cylindrical plexiglas rods shows clearly some of the potentially destructive stresses which develop.

When a small electric detonator is exploded at or near the center of the end of a rod, the detonator produces a transient essentially sawtooth stress disturbance of about 3 microseconds duration, corresponding to a wavelength of 7 mm, in plexiglas. Wave reflection and interference patterns are controlled by shaping the end of the rod on which the detonator is exploded. An interesting specific case is where the end of the rod is machined to a truncated cone shape, with the various cone angles ranging from 45º to 130º.

Three distinct fracture systems can be recognized. The most common one is the fractures which occur along the axis about one rod diameter from the explosively loaded end. This group of fractures is generated by the reflected tensile wave coming in from the boundary of the cylinder as shown in Fig. 12-33. The reflected wave which is cylindrically convergent and focused along the axis of the rod, produces high tensile stresses along the axis, except above the point A where the fractures begin. It is geometrically impossible for reflected tensile waves to reach into this region. The distance OA depends upon cone angle and rod diameter and can be readily computed. For specimens having cone angles in the range 80º to 90º, the fractures extend only a short distance along the axis, elongating downward as the cone angle increases.

When the detonator is shifted from its central position, the same type of fractures appear, but they are no longer on the axis of the specimen. They now lie along a line parallel to the axis and symmetrically placed with respect to the vertical passing through O, the point of application of the detonator (see Fig. 12-34).

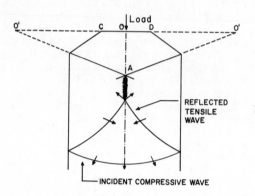

Fig. 12-33. Wave interactions leading to fracturing in a truncated cylindrical rod loaded at the center of one end.

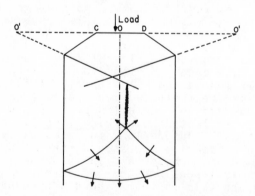

Fig. 12-34. Similar to Fig. 12-33 except load is asymmetrically placed.

In both cases, fracturing stops because of the decreasing obliquity of the elements of the front of the converging tensile wave at the instant of its complete collapse.

For a given rod diameter and a given cone angle,

it is possible to vary the length of the fracture
region AB (Fig. 12-35) by varying the diameter CD of
the upper face of the truncated cone.

   A second system of fractures composed of hori-
zontal cracks is located above the lower system just
described. These are just barely visible in most
specimens. The shear wave generated through reflection at the boundary of the cylinder in the manner
illustrated in Fig. 12-35, initiates these cracks.
This system of shear fractures will only be observed
when the shape of the specimen places the front of
final convergence .of the shear wave A' above A, the
point of convergence of the tensile wave. When A' is
located below A, the second system of fractures will
be concealed by the first.

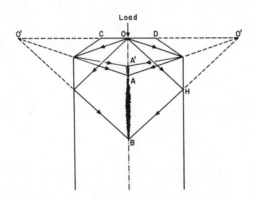

Fig. 12-35.   Ray diagram showing how shear waves
              cause horizontal fractures to occur
              between A and A'.

   In small cone angle specimens, angles in the
neighborhood of 45º, there is a third system of
fractures, clearly visible in specimens. These are
due to the generation of a tensile wave through
reflection of the incident compression wave from the
surface of the truncated cone.

   In a somewhat similar experiment, a solid cylinder of explosive having the same diameter as the rod,
was placed on the end of the rod and exploded. While
both this experiment and the previous point loading
one involve explosive charges, they simulate the
results of impact loading as well. The pattern of

waves that will develop has been described in Chapter 8. In the left hand drawing of Fig. 12-36, the detonation front is shown moving down through the explosive, reaching the end MN of the solid cylinder and subjecting it suddenly to an intense compressive load.

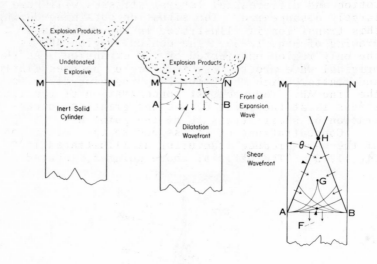

Fig. 12-36.   Pattern of waves developed when a
              cylinder of explosive is detonated on
              the end of an inert solid cylinder.

A sharp fronted pure dilatation compressive wave, AB in the middle drawing, is set up which moves down the cylinder with dilatation velocity c_1. At the same time, the expansion rapidly works its way inward, establishing a tensile release wave. This is the state of affairs at the instant the expansion has progressed about half way through the cylinder. The expansion wave moves with the velocity c_1 of the dilatation waves and converges toward the axis, amplifying tremendously the magnitude of the expansive tensile stresses. Theoretically, these stresses could build up to an infinite value along the axis. Because the wave does not have a true coherent front in the sense that a Huygen's envelope can be constructed, the problem cannot be treated analytically. Each

wavelet can be presumed to have no energy but their
combined effect is to generate a kind of wave expansion having its highest stress gradients when moving
across the end section MN with gradients lessening as
convergence occurs further and further along the
cylinder. A few cylinder diameters down the cylinder,
its entire cross section is in more or less uniform
motion and differential lateral stresses will have
largely disappeared. The situation obtaining during
this transition is illustrated in the right hand
drawing of Fig. 12-36. The conical region AFB is
the only region unaffected by the expansion. As the
original wave progresses, this region becomes flatter
and flatter, with the point F approaching close to
the line AB. The greatest concentration of tensile
stress is at the point G, and the greatest concentration of shear stress is at the point H.

Concentrations of stress due to the reflection
of the wave produce fracturing as illustrated in
Fig. 12-37. The material above point A, located

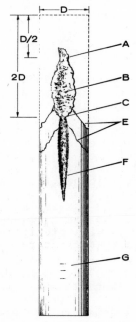

Fig. 12-37. Fracturing generated by wave interactions
shown in Fig. 12-36.

about one cylinder radius from the original end of
the cylinder, is completely blown away and is unrecov-
erable. This point is at the upper terminus of a
recoverable teardrop-shaped region which is heavily
internally fractured. It forms a cohesive plug which
separates completely from the rest of the cylinder.
The fracturing is layered and forms a rough chevron
pattern symmetrical around the axis of the cylinder.
The stem of the plug, nearest A, is completely fused,
indicating that it was heated above the melting
point during passage of the wave.

The diameter of the teardrop region or plug
increases, reaching a maximum at B of about one
cylinder radius. This is about half way through its
length, at which point its diameter begins decreasing, becomes zero, and terminates at C, about two
cylinder diameters from the top end of the cylinder.

A few large fragments, E, which break off around
the base of the plug are generally recoverable.

A more mildly fractured central region, F in the
figure, lies below the plug, extending down one or
two cylinder diameters. This region consists of
numerous preferentially oriented microfractures
nestling relatively close to the axis.

The fractures at G are spall fractures resulting from the reflection of the wave at the end of
the rod.

Using high-speed photographic techniques, it is
possible to follow the progress of the fracturing in
plexiglas. It is found that the central fractures
carve out a plug (region A to C in Fig. 12-37) as a
result of convergence of the longitudinal wave,
whereas the fractured region (C to F) which lies
further down the cylinder is caused by convergence of
the shear wave. The two distinctly unlike origins of
fracture cause a striking difference in the appearance
of the two regions.

Similar fracturing occurs in cylindrical metallic rods. Onset of the fractures cannot be photo-
graphed internally as with plexiglas, but sectioning
shows that the resultant fracture pattern is
essentially the same.

Chapter 13

SPALL FRACTURING

INTRODUCTION

While making scientific investigations on the
effect of explosives, Hopkinson, in 1914, observed
a very interesting phenomenon which has since been
termed a Hopkinson fracture, scabbing, or spalling.
Spalling is fracturing caused when a high intensity,
transient stress wave reflects from a free surface.
The wave may be generated, as Hopkinson did, through
the detonation of an explosive charge placed in intimate contact with the material at some distance from
the free surface where spalling occurs, by hypervelocity impact, or by the sudden application of
energy to a surface such as fast absorption of X rays
originating in a nearby nuclear blast. Spectacular
results of spalling are the havoc created by the
fragments hurled from the interior walls of tanks and
missiles when explosive charges are plastered on
their outer walls, the breaching of concrete barri-
cades by large, high speed projectiles, the creation
of a towering plume of water when an explosive is
detonated below the surface, or the hurtling of fly
rock associated with large blasts in rock.

There was a hiatus of about thirty-five years
between Hopkinson's work and the resumption of
serious scientific study of spalling phenomena. Beginning about 1950, activity in this field increased
and spalling has become an especially important,
significant, and active area for study by a large
number of investigators working on a diversity of
problems. The knowledge gained has been invaluable
in the solution of many design problems, whether it
is the prevention of fracturing that is of prime
concern, or, equally important, the generation of

preselected fracturing.

Material properties such as fracture strength
and mechanical anistropy will control in a large
measure the character of the spall process. The
plume of water results from the cohesionless character of water whereas the large ponderable fragments
of a missile housing struck by a meteorite results
from the great resistance of metal to fracturing.

DYNAMICS OF CREATION OF SPALL

Spalling is the direct consequence of interference near a free surface between the portion of an
oncoming incident compression wave which has not yet
been reflected and the portion which has been reflected and transformed into a tensile wave.

When a compressive longitudinal wave strikes a
free surface normally, the boundary conditions, continuity of stress and continuity of particle velocity, will be preserved only if the wave is reflected
as a tensile longitudinal wave of equal strength.
Oblique incidence reflection is more complex but
generally a tensile wave somewhat lesser in magnitude
will be generated. When the portion of the wave
which has not yet reached the surface and the portion that has been reflected interfere with one another, there results a distribution of stress conduc-
ive to the generation of a fracture. Usually the
amount of tension increases as the reflected wave
moves back inward from the surface.

An example of this superposition of waves is
illustrated in Fig. 13-1 for a triangular or sawtooth pulse of maximum intensity σ_o and length Λ,
where it is assumed that half of the wave has been
reflected. As the wave continues to reflect and
move inward to the left, the resultant tension CD
increases. Finally, if the material does not break
first, the tension reaches a maximum value σ_o.

The maximum stress that can ever be attained at
each particular point of Fig. 13-1 is plotted in
Fig. 13-2 as a function of position in the material
for points near the surface. The pattern of distribution of maximum stress is critically dependent upon
the shape of the stress wave, each wave type having
its own pattern. Distributions for two other wave
types, a flattopped wave and a square wave, are

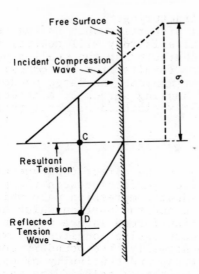

Fig. 13-1. Superposition of stresses during reflection of a sawtooth wave from a free surface.

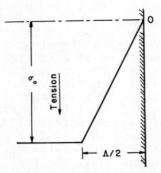

Fig. 13-2. Maximum stress attained during reflection of a sawtooth wave from a free surface as a function of distance in from the surface. Note stress at surface is always zero with tensile stress increasing to a maximum at distance $\Lambda/2$ from surface.

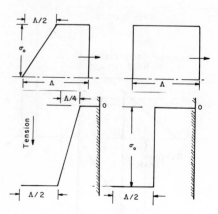

Fig. 13-3.   Plots similar to Fig. 13-2 for
             flattopped wave (left) and square
             wave (right).

shown in Fig. 13-3. Note that for all wave types
the stress at the free surface is zero, which is as
it must be.
    In the case of the sawtooth wave, the tensile
stress increases linearly to a maximum value of σ_0,
in a distance $\Lambda/2$, the rate of increase being twice
as great as it was in the original wave. The maximum stress then remains constant at a level of σ_0
as the wave moves back through the material. The
flattopped wave must move back into the material a
distance equal to half the length of its flat portion, $\Lambda/4$, before the tensile stress begins to
increase at all, after which the pattern of the
stress increase is similar to that of the sawtooth
wave. The stress in the case of the square wave
changes discontinuously from zero to maximum stress
at a distance from the surface equal to half the
length of the wave, $\Lambda/2$. Thus for the cases of the
flattopped and square waves, maximum stress is
reached much further in from the free surface than
for the sawtooth wave.
    The plane wave striking a surface normally is
an unusual case. Generally the wave front is divergent and strikes obliquely. These cases are treated
in later sections of this chapter. Although they
are more complicated, it is not too unrealistic to

treat a divergent wave as if its front were made up
of a series of straight segments and, further, to
neglect the presence of the shear wave generated during oblique reflection.

If the material is capable of withstanding the
tensile stresses that develop within it, the reflected wave will move unimpeded without alteration
through the material. If not, it will break, further
affecting the distribution of stress and the progress
of the waves. Whether spalling occurs at all, and
where the fracture or fractures are located depend
upon three factors: the resistance of the material
to fracture; the magnitude of stress in the stress
wave; and even more important, the shape of the
stress wave.

SPALLING OF STRONG MATERIALS

Spalling in metals, which are homogeneous,
isotropic materials, about equally strong in tension
and compression, has been studied extensively and is
well understood. Waves of relatively weak intensity
produce in mild steel a single, well defined fracture. If the stress is high, a group of several
parallel fractures may form.

The single spall obviously occurs when the
stress developed through interference of the incident
and reflected portions of the transient stress pulse
reaches a value higher than the mild steel can with-
stand. This critical stress level is called the
critical normal fracture strength of the material.
Values determined of the dynamic tensile strength of
various metals are listed in Table 13-1. In general,

Table 13-1. Critical normal fracture strength

Metal	Critical normal fracture strength (kilobars)
4130 steel	30.0
Copper (annealed)	28.2
Brass	21.0
1020 steel (annealed)	12.3
24S-T4 aluminum	9.5

they are considerably higher than the tensile
strengths which are determined statically. Unfor-
tunately, it is not easy to specify precisely either
the rate of strain or the complete state of stress
at which the dynamic tensile strength or critical
normal fracture strength is being measured. For all
practical purposes, the rate of strain is infinite,
the stress at the point of fracture changing
instantaneously and discontinuously from one of
hydrostatic compression to one of hydrostatic tension. Nonetheless, the critical normal fracture
strength, representing as it does the resistance of
the material to a dynamic tensile load, is an important mechanical property of any material.

Multiple spalling, the development of several
parallel, juxtaposed fractures, occurs when the level
of the stress wave becomes more than twice the
critical normal fracture strength of the material.
Single spalling is merely a special case of multiple
spalling for the situation in which the maximum intensity of the transient stress is greater than the
critical normal fracture strength of the material and
less than twice its critical normal fracture
strength.

When several fractures develop because of the
high intensity of the stress wave, the first spall
fracture, upon its abrupt creation, places a free
surface in front of the trailing portion of the
original wave which is immediately effective in reflecting the remainder of the wave, that portion of
the wave which has not been trapped in the first
spall. This process continues until the intensity
of the rear portion of the wave is no longer greater
than the critical normal fracture strength of the
material. A bell shaped wave has been used in Fig.
13-4 to illustrate the process.

The drawings in the figure also illustrate the
dependence of spall thickness, or the distance
between fractures, on wave shape. The bell shaped
wave, which is characteristic of explosive loading,
initially decays slowly, then rapidly, and finally
tails off gradually, generating a first spall that
is thick, a second spall that is thin, and a final
spall that is likely to be thick. Experiments confirm this prediction. The variation in thickness
depends upon the fact that a spall will only occur
when the critical normal fracture stress is reached
and this can only be reached when sufficient tension

can be achieved through superposition of the incident
and reflected waves.

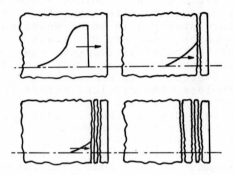

Fig. 13-4. Mechanics of generation of multiple
                       spalls.

        In most real situations, the initially generated
stress pulse changes its shape and intensity as it
moves through the material. A high intensity, initially short duration pulse usually decays while at
the same time becoming longer. Lengthening of the
pulse flattens it. A consequence of this flatten-
ing is that, on reflection, the pulse must go further into the material to develop the critical
fracture stress than was necessary when decay of
stress behind the front of the pulse was more rapid.
An example of such pulse lengthening is illustrated
in Fig. 13-5 in which spall thickness is plotted as
a function of plate thickness for some explosively
loaded steel plates. In general, spall thickness
increases with increasing plate thickness. Note
that there is a cut off point beyond which the intensity of the stress is too small to generate a spall
at all.
        Since stress and particle velocity are linearly
related, the higher the stress, the greater will be
the velocity with which the material tears loose and
moves away. At low stress levels, the spall fracture
may be the only fracture which develops and in fact
often remains invisible within the material. At
higher stress levels, the spalled portion will likely have sufficient velocity to tear itself loose

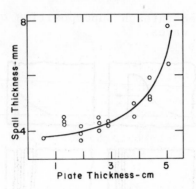

Fig. 13-5.   Spall thickness as a function of plate
             thickness for an explosively loaded
             steel plate. A layer of explosive
             6.4 mm thick was placed on front
             surface of plate.

from the parent body and hurl itself with considerable velocity through the air. In multiple spalling
a velocity gradient develops among the spalls, and
the relative displacements of the several spalls in
the plate differ.
     The respective velocities of spalls can be cal-
culated if the shape of the stress wave is known.
Consider the bell shaped wave of Fig. 13-4 and
assume that stress σ as a function of time t is
given by $\sigma(t)$ and that the density of the material
is ρ. Then the velocity V_1 of the first spall will
be given by

$$V_1 = (1/2\rho L_1)\int_0^{2L_1/c_1} \sigma(t)dt \qquad (13.1)$$

where L_1 is the thickness of the first spall. Since
spalling occurs when the tensile stress reaches the
critical normal fracture strength of the material,
σ_c, $2L_1$ is the distance along the wave needed to
reduce the intensity of stress to a value just σ_c
less than the maximum stress σ_0. The second spall
forms when the stress is reduced again by the same
increment σ_c and the velocity V_2 of the second
spall will be given by

$$V_2 = (1/2\rho L_2) \int_{2L_1/c_1}^{2L_2/c_1} \sigma(t)dt. \qquad (13.2)$$

The velocities of succeeding spalls can be calculated in the same manner. If the small explosive charge which is used to spall a steel plate generates a stress near the spall surface of about 7 kilobars, the velocity of the outermost spall will be in the neighborhood of 200 m/sec.

SPALLING IN LAMINATED MATERIALS

The structure of the material can influence profoundly the pattern of spalling. Many materials in which spalls occur are not homogeneous or isotropic. A simple type of a nonhomogeneous structure to consider is that in which the material is layered, each layer being of the same material but separated by planes of weakness. Many rocks have such a layered structure and some metals, particularly dirty, strongly textured, heavily rolled steel plate. An idealized layered structure is a group of plates bundled together so that each plate is in intimate contact with but not cemented to its neighbor. When these layers lie perpendicular to the direction of propagation of the stress wave, the stress wave, being one of compression, crosses these boundaries without diminution or change in form. On reflection, however, with the wave changed to tension, the weak boundaries can no longer support it so that the layers peel off one after the other in the same way multiple spalling occurs, except that the location of the spalls, the boundaries themselves, are preselected. The velocity with which each plate flies off can be calculated in the same way that the velocity of a spall was calculated.

It should be noted that if the layers lie parallel to the direction of propagation of the wave, the spall pattern is relatively unaffected by the layering and the spalling occurs at essentially the same location it would in a homogeneous material as illustrated in Fig. 13-6.

The situation becomes more complicated when the

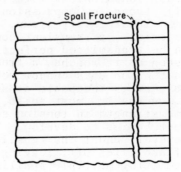

Fig. 13-6.   Spall fracture generated in laminated
             material when laminations are
             perpendicular to free surface.

layers consist of dissimilar materials, for when a
longitudinal stress wave strikes a boundary between
two dissimilar materials, part of the stress will be
transmitted and part will be reflected (see Fig.
7-7). The partitioning of stress can be calculated
from Eqs. (7.5 and 7.6).   The significant case
for spalling is the one in which $\rho c_1 > \rho c_1'$, where a
compressive wave will be reflected as a tension
wave that can produce spalling.
     It is tedious but entirely practicable to ana-
lyze in detail what will happen in a body containing
many laminations of several different materials. A
simple situation to view is that in which a single,
heavily spalled plate is backed up with a plate of
some different material. When this is done, the
spall pattern is changed. For example, a steel
plate backed up with an aluminum plate reflects
only about 50 per cent of the wave and as a result
the number of spall fractures is much less. In
practice, undesirable spalling can often be elimi-
nated by using backing plates or other momentum
traps for bleeding the wave out of the parent body.

          SPALLING IN COHESIONLESS MATERIALS

     Spalling is not limited to strong and rigid
materials, but occurs frequently in soils, powders,

liquids, and other cohesionless materials. Rock,
being relatively strong in compression but weak in
tension, forms an intermediary group. In incompetent
cohesionless materials, a transient compression
stress wave can be transmitted perfectly satisfac-
torily but as soon as it reaches a free surface and
is reflected, a tension wave cannot develop simply
because the material is incapable of supporting a
tensile stress. Actually, most soils, powders, and
fluids have some strength in tension so that a reflected wave just begins to develop. The material
soon breaks, however, under the tensile load and a
thin spall or flake flies off, leaving a new free
surface and permitting multiple spalling to develop.
Flake after flake leaves the surface, each bit ab-
sorbing a small portion of the momentum of the
incident wave. The velocity V_f of any one of the
small flakes is given by

$$V_f = 2\sigma/\rho c_1 \qquad (13.3)$$

where σ is the value of the stress at the front of
the incident wave at the time the flake leaves the

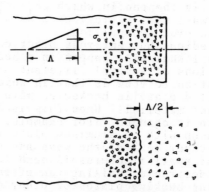

Fig. 13-7. Flaking by spalling in cohesionless
                      material.

surface, and ρ and c_1 are, respectively, the density
and longitudinal wave velocity of the material. The
flakes having the highest velocity are those thrown
off first and their velocity will be equal to

$2\sigma_o/\rho c_1$ where σ_o is the maximum stress level of the
wave. For a sawtooth wave having a wave length Λ,
the material will continue to flake off back to a
distance $\Lambda/2$ from the original surface, as shown
in Fig. 13-7. By the time the reflected wave has
traveled a distance $\Lambda/2$, all of the momentum of the
wave has been taken up by the flakes and no portion
of it returns into the material. The flakes themselves appear as a cloud of debris, the rear portion
of which is stationary with the front moving forward
with a velocity of $2v_o$, where v_o is the maximum wave
particle velocity. Such a cloud is an extremely
effective shock wave attenuator, limited, of course,
to one time usage.

Spalling in rocks frequently manifests itself
as a series of thin layers, or multiple spalls.
Rock types differ greatly and some rocks, such as
trap rock in which the compressive and tensile
strengths are about the same, behave as metals. With
others, such as granite, where the ratio between
compressive and tensile strength may be as high as
70 or 80, flaking or multiple spalling is common
because while the granite can transmit high compressive stresses without failing, it fails easily
under the strong tensile stresses generated on reflection of the compressive waves.

POSITION OF SPALL FRACTURE SURFACE AS A FUNCTION OF OBLIQUITY OF WAVE FRONT

The position of the spall with respect to the
free surface, for an incident wave of specified
shape and intensity, will depend upon the angle the
incident wave front makes with the free surface
that it strikes.

Assume that the incident wave is sawtooth, of
length Λ, and maximum intensity $2\sigma_c$ at its sharp
front, and that the material is plexiglas with a
Poisson's ratio of 0.40. Consider four angles of
incidence: 0°, 30°, 45°, and 60°. For 0°, the
thickness of the spall will be 0.25Λ since the reflected wave is not reduced in intensity. At other
angles of incidence, the intensity will be reduced
by the factor R, where R, the reflection coefficient discussed in Chapter 7, is given by

R = (tan βtan² 2β - tan γ)/(tan βtan² 2β + tan γ).
$$(13.4)$$

For 30°, the reduction factor is 0.89; for 45°, 0.79;
and for 60°, 0.73.

The physical location of the macrofracture that
constitutes the spall will be fixed at that place
where the principal stress, σ, developed by the
interference or superposition of the front of the
reflected wave and the rear portion of the incident
wave reaches the critical value σ_c. To find this
location, it is necessary to determine the stress,
designated σ_1 in the incident wave at which the
resultant stress is σ_c. The stress at the front of
the reflected wave in the present example will be
$2R\sigma_c$. The stress σ_1 which will lead to fracturing
can be found by solving the following equation
derived in Chapter 6:

$$\sigma_c = [1/2(1 - \nu)]\left\{2R\sigma_c - \sigma_1 + \right.$$
$$+ (1 - 2\nu)[(2R\sigma_c)^2 + \sigma_1^2 - 4R\sigma_c\sigma_1 \cos 4\alpha]^{\frac{1}{2}}\right\}.$$
$$(13.5)$$

This equation involves R and the angle α between the
two interfering wave fronts. The angle α depends
on the angle of incidence γ, γ in fact being equal
to 2α.

Once σ_1 has been determined, the determination
of the position of the spall for the linearly decaying wave considered here is geometrically straightforward. The distance x behind the incident wave
front at which the stress has been reduced to σ_1
is given by

$$x = [1 - (\sigma_1/2\sigma_c)]\Lambda.   (13.6)$$

Noting that the wave front is inclined to the free
surface at the angle of incidence γ, the distance D
from the free surface to the spall can be determined
geometrically.

The situation for a 30° angle of incidence is
illustrated in Fig. 13-8. In this case, x is equal
to 0.47Λ and D is equal to 0.27Λ.

The distance of the spall from the free surface
is plotted as a function of angle of incidence in
Fig. 13-9. Generally the distance increases with

increasing angle of incidence, slowly up to about
30°, and then rapidly.

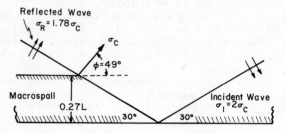

Fig. 13-8.   Position of spall and orientation of
              principal stress causing spall for a
              30° incident sawtooth wave of
              intensity $2\sigma_c$. ν = 0.40.

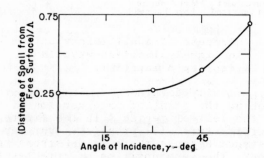

Fig. 13-9.   Distance of spall from free surface
              divided by wave length as a function
              of angle of incidence for the saw-
              tooth wave of Fig. 13-8.

          NATURE OF SPALL FRACTURE SURFACE AS FUNCTION
          OF OBLIQUITY

     The direction of the highest principal stress
developed by the intersecting waves determines the
orientation of each of the microfractures that
combine to form the spall in a mechanically isotropic material. Each microfracture is perpendicu-

lar to the principal stress that produces it so that
the character of the macrofracture of the spall
surface will depend on obliquity.

The process of spall formation in such a material is illustrated in Fig. 13-10 for an obliquely
incident plane wave. As the two waves, incident
and reflected, interfere and a principal stress

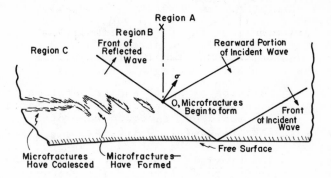

Fig. 13-10. Process of spall formation for an
            obliquely incident wave in an
            isotropic material.

equal to σ_c finally develops, a microfracture opens
up just behind the point of intersection of the two
waves, to the left of region A in the figure. No
single microfracture will propagate over any appre-
ciable distance because there is no stress to sustain
propagation. The superposition of stresses that
causes microfracturing is a highly localized and
transient event, as was the case in the development
of corner fractures. The point or line of action, O,
moves quickly across the surface over which the
macro spall finally forms. The microfractures
(region B in Fig. 13-10), left in the wake of the
progressing intersection, join together through
further fracturing as the spall pulls loose (region
C in the figure). The driving force which pulls
the spall loose is the momentum trapped in the region
lying below the macro spall fracture surface.

The surface of such a macro spall fracture will
appear extremely flaky as can be observed in metals
where the spalls have been produced by obliquely
striking waves. Under some conditions, the trapped

momentum is insufficient to pull the spall loose
although the interfering stress waves have left a
fully developed array of microfractures. Such
arrays occur in metals, rocks, soils, and plastics
that have been explosively attacked. They can be
easily seen near free surfaces in the transparent
plastic plexiglas.

As described in Chapter 6, the angle ϕ between
the highest principal stress σ and the bisector OX
(Fig. 13-10) between the two interfering wave fronts
is related to the relative stresses and the angle
2α included between the two fronts. Specifically,
for a saw tooth wave of length L and maximum intensity $2\sigma_c$, ϕ is given by

$$\tan 2\phi = \left[(R\sigma_c - \sigma_1)/(R\sigma_c + \sigma_1) \right] \tan 2\alpha. \quad (13.7)$$

Substitution of appropriate values of σ_1 and R for
angles of incidence of $0°$, $30°$, $45°$, and $60°$ gives
values of ϕ for plexiglas ($\nu = 0.4$) equal to $90°$,
$49°$, $45°$, and $39°$, respectively. The respective
orientations of the fractures that would develop at
each of these four angles of incidence are illustrated in Fig. 13-11.

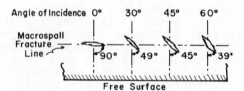

Fig. 13-11.   Orientation of microfractures for
              four angles of incidence. $\nu = 0.40$.

In a mechanically anisotropic material, a
factor of directional weakness is added to stress
level to control the orientation of the microfractures. The microfractures often tend to lie perpendicular to the direction in which the material is
weakest but sufficient stress must develop in that
direction to pull the material apart, a condition
which may not always obtain. However, in an anisotropic material, the orientation of the microfractures can be anywhere between the direction of great-

est strength and the direction of greatest weakness,
depending upon the configuration of the stress wave
interference pattern, the intensities of the stresses
involved, and the directional dependence of strength.

SPALLING PRODUCED BY A SPHERICAL WAVE

Point loadings such as high speed impacts and
concentrated explosions produce spherically expanding wave fronts. While at large distances these
fronts can be treated as if they were plane, when
the free surface is close to the explosion or point
of impact, the curvature of the wave front has a
strong influence on the spall pattern. One effect
of the curvature is illustrated in Fig. 13-12.

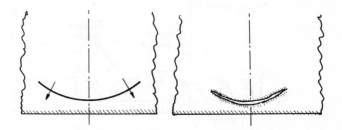

Fig. 13-12. Curved macro spall fracture produced
             by a spherical wave striking a free
             surface.

Along the perpendicular drawn from the origin of the
shock to the free surface, the incident and reflected
waves are parallel, but on each side of this perpendicular, the waves are inclined to one another, the
inclination depending upon the distance off axis
and the curvature of the incident wave front. Due
to this oblique incidence, the intensity of the
reflected wave will vary along the wave front, generally decreasing with increasing distance from the
axis. In addition, the intensity is further reduced
by divergence of the wave. The net effect, shown in
the right drawing of Fig. 13-12, is that the thickness of the spall increases as the distance from the
axis increases, finally stopping altogether when the
reflected wave becomes too weak.

Wave front curvature will also influence the microfracturing that accompanies spall formation. As off axis the reflected and incident waves are not parallel, the microfractures which form will no longer be parallel to the free surface. The inclination of each microfracture can be computed from the general relationship

$$\tan 2\phi = \left[(\sigma_1 - \sigma_2)/(\sigma_1 + \sigma_2)\right] \tan 2\alpha \qquad (13.8)$$

discussed in Chapter 6. Its position can be located by applying Eq. (13.5) to find where the critical normal fracture stress develops.

The results of a detailed analysis of a case of oblique angle spalling are illustrated in Fig. 13-13.

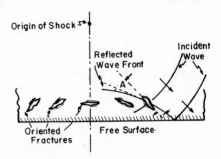

Fig. 13-13.   Orientations of microfractures produced by spherical wave of Fig. 13-12.

The angle A that the microfracture makes with the reflected wave front which is equal to $\alpha - \phi$, is zero along the axis. Off axis, the angle A at first increases and then diminishes to zero, resulting in the pattern of microfractures shown in the figure.

APPENDIX

BIBLIOGRAPHY

INTRODUCTION

Specific topics are not referenced in STRESS
TRANSIENTS IN SOLIDS since it is primarily a class-
room text. It is, of course, based on many books
and technical articles which provide excellent
supplemental reading for fuller development of basic
principles and greater appreciation of the further
application of these principles. The several books
and articles listed below have been chosen on the
basis of historical interest, engineering elegance,
and contemporary importance. The books are briefly
annotated.

BOOKS

Chou, P. C. and A. K. Hopkins, eds., 1972,
DYNAMIC RESPONSE OF MATERIALS TO INTENSE IMPULSIVE
LOADING, U. S. Air Materials Laboratory, Wright
Patterson AFB, Ohio, 555 pp.
Contains articles especially prepared by nine
experts in the field of transient stress loading.
Its central theme is the response of materials to
microsecond intense stress loading. STRESS TRANSIENTS
IN SOLIDS provides an excellent foundation on which
to begin the study of this.

Davids, Norman, ed., 1960, STRESS WAVE PROPAGA-
TION IN MATERIALS, Interscience, New York, 337 pp.
Contains the fifteen papers that were presented
at an international symposium, attended by about
two hundred engineers and scientists active in stress
wave research and held at The Pennsylvania State
University in 1959.

Ewing, W. M., W. S. Jardetsky, and Frank Press, 1957, ELASTIC WAVES IN LAYERED MEDIA, McGraw-Hill, New York, 380 pp.
Complete treatise on this subject. Suitable for a text in advanced seismology. STRESS TRANSIENTS IN SOLIDS provides good physical insight into some of the problems discussed with mathematical rigor in this book.

French, B. M. and N. M. Short, eds., 1968, SHOCK METAMORPHISM OF NATURAL MATERIALS, Mono Book Corp., Baltimore, 644 pp.
Proceedings of a NASA sponsored conference held at Greenbelt, Md., 1966. Contains forty-three papers emphasizing the effects of transient stresses on rocks and minerals and how these can be recognized. This field is one of the most exciting new developments in geology. Topics include behavior of solids under shock waves, phase changes produced by shock pressures, microscopic deformational effects of shock waves from nuclear and chemical explosions, effects of heterogeneous media on shock wave propagation, and criteria for recognizing ancient meteorite impact.

Goldsmith, Werner, 1960, IMPACT, Edward Arnold, London, 379 pp.
Treats the dynamics and mechanics of physical impact between solid bodies. Impacts are those lying between the domain of static loading and that of fluid behavior of materials which is most readily treated by methods of hydrodynamics. The stress levels considered lie within two orders of magnitude of either side of the yield stress. Very useful reference source.

Jaeger, J. C., 1956, ELASTICITY, FRACTURE AND FLOW WITH ENGINEERING AND GEOLOGICAL APPLICATIONS, Methuen, London, 152 pp.
Excellent discussion of the interrelationships among static stress and strain both elastic and inelastic. Good comparative reading.

Kinslow, Ray, ed., 1970, HIGH-VELOCITY IMPACT PHENOMENA, Academic Press, New York, 579 pp.
Compendium of invited papers by experts that in many instances show applications of the fundamentals of transient stress phenomena to practical engin-

eering problems. Excellent supplementary reading.

   Kolsky, Harry, 1953, STRESS WAVES IN SOLIDS,
Oxford University Press, London (Dover, New York,
1963), 213 pp.
   The classic on stress waves. Surveys the
theoretical core of knowledge on the subject and, in
addition, correlates this knowledge with experimental
results. It covers stress waves in both elastic and
imperfectly elastic media. It should be studied by
every serious student of stress wave phenomena.

   Lee, G. H.; 1950, AN INTRODUCTION TO EXPERIMENTAL
STRESS ANALYSIS, Wiley and Sons, New York, 319 pp.
   One of the many elementary texts on the subject.
An especially lucid and concise book.

   Love, A. E. H., 1944, THE MATHEMATICAL THEORY
OF ELASTICITY, Dover, New York, 643 pp.
   The classic work on elasticity.

   Rinehart, J. S., 1960, ON FRACTURES CAUSED BY
EXPLOSIONS AND IMPACTS, Quarterly Colorado School of
Mines, Golden, Colorado, 155 pp.
   An adaptation of an instruction manual prepared
for the U. S. Air Force to provide background material for important defense problems, especially
spalling. Discusses the physical basis of transient
stress wave interactions and how materials respond
to them

   Rinehart, J. S. and John Pearson, 1954, BEHAVIOR
OF METALS UNDER IMPULSIVE LOADS, American Society for
Metals, Cleveland, Ohio (Dover, New York, 1964),
256 pp.
   Forerunner to STRESS TRANSIENTS IN SOLIDS but
with more emphasis on material behavior. Mathematical and geometric formulations are much less
developed. Contains comprehensive bibliography up
to 1953.

   Rinehart, J. S. and John Pearson, 1963,
EXPLOSIVE WORKING OF METALS, Pergamon, New York and
London, 351 pp.
   Outlines the fundamental principles that provide
the basis for using explosives to work metals and
then discusses engineering fundamentals and practices.
Excellent supplementary reading.

22   Shewmon, P. G. and V. F. Zackay, eds., 1961,
RESPONSE OF METALS TO HIGH VELOCITY DEFORMATION,
Interscience, New York, 491 pp.
     Proceedings of a conference held on this subject
at Estes Park, Colorado, in July 1960. Many of the
phenomena discussed can be understood in the light of
the principles set forth in STRESS TRANSIENTS IN
SOLIDS.

     Timoshenko, S. and J. N. Goodier, 1951, THEORY
OF ELASTICITY, McGraw-Hill, New York, 606 pp.
     Complete treatise on static elasticity. The
analytical solutions to many problems are worked out.

     Wasley, R. J., 1973, STRESS WAVE PROPAGATION IN
SOLIDS, Marcel Dekker, New York, 279 pp.
     The most recent book on stress wave propagation,
a good general introduction to the subject. Discusses
the fundamentals of elastic wave propagation and
suggests alternate approaches to cases where elastic
wave propagation does not adequately describe physical
response. It contains considerable background information for the study of nonelastic wave phenomena.

TECHNICAL ARTICLES

     Blake, F. G., Jr., 1952, "Spherical wave propa-
gation in solid media", J. Acoustical Soc. Amer.,
24, 211.

     Broberg, K. B., 1956, "Shock waves in elastic
and elastic-plastic media", Kungl. Fortif. Befast.
Forsk. och Forsok., Stockholm, Report No. 109, 12.

     Davies, R. M., 1948, "A critical study of the
Hopkinson pressure bar", Roy. Soc. Phil. Trans.,
A240, 325.

     Davies, R. M., 1956, "Stress waves in solids",
in SURVEYS IN MECHANICS, ed. by G. K. Batchelor and
R. M. Davies, p. 64, Cambridge University Press,
Cambridge, England.

     Duvall, G. E., 1961, "Some properties and
applications of shock waves", in RESPONSE OF METALS
TO HIGH VELOCITY DEFORMATION, ed. by P. G. Shewmon
and V. F. Zackay, p. 165, Interscience, New York.

Goldsmith, Werner and W. A. Allen, 1955, "Graphical representation of the spherical propagation of explosive pulses in elastic media", J. Acoustical Soc. Amer., 27, 47.

Hopkinson, Bertram, 1905, "The effects of momentary stresses in metals", Proc. Roy. Soc., London, 74 (see also his SCIENTIFIC PAPERS, Cambridge University Press, 1921, p. 49).

Hopkinson, Bertram, 1914, "A method of measuring the pressure produced in the detonation of high explosives or by the impact of bullets", Roy. Soc. Phil. Trans., A213, 437.

Hughes, D. S., W. L. Pondrom, and R. L. Mims, 1949, "Transmission of elastic pulses in metal rods", Phys. Rev., 75, 1552.

Kinslow, Ray, 1963, "Properties of spherical stress waves produced by hypervelocity impact", Arnold Engineering Development Center, Arnold Air Force Station, Tenn., AEDC-TDR-63-197, 44 pp.

Kolsky, Harry, 1959, "Fractures produced by stress waves", in FRACTURE, ed. by B. L. Auerbach, D. K. Felbeck, G. T. Hahn, and D. A. Thomas, p. 533, Technology Press and Wiley, New York.

Rice, M. H., R. G. McQueen, and J. M. Walsh, 1958, "Compression of solids by strong shock waves", in SOLID STATE PHYSICS, Vol. 6, ed. by Frederick Seitz and David Turnbull, p. 1, Academic Press, New York.

Rinehart, J. S., 1965, "Compilation of dynamic equation of state data for solids and liquids", U. S. Naval Ordnance Test Station, California, Report NOTS-TP-3798, 330 pp.

Roesler, F. C., 1955, "Glancing angle reflection of elastic waves from a free boundary", Phil. Mag., 46, 517.

Schardin, V. H., 1950, "Ergebnisse der Kinematographischen Untersuchung des Glasbruchvorganges", Glastechn. Ber., 23, 1, 67, and 325.

Selberg, H. L., 1952, "Transient compression waves from spherical and cylindrical cavities", Arkiv för Fysik, 5, 97.

Sharpe, J. H., 1942, "The production of elastic waves by explosion pressures, I. Theory and empirical field observations", Geophysics, 7, 144.

INDEX

191818
192183